National Forest Inventories: Contributions to Forest Biodiversity Assessments

Managing Forest Ecosystems

VOLUME 20

Series Editors:

Klaus von Gadow

Georg-August-University,
Göttingen, Germany

Timo Pukkala

University of Joensuu,
Joensuu, Finland

and

Margarida Tomé

Instituto Superior de Agronomía,
Lisbon, Portugal

Aims & Scope:

Well-managed forests and woodlands are a renewable resource, producing essential raw material with minimum waste and energy use. Rich in habitat and species diversity, forests may contribute to increased ecosystem stability. They can absorb the effects of unwanted deposition and other disturbances and protect neighbouring ecosystems by maintaining stable nutrient and energy cycles and by preventing soil degradation and erosion. They provide much-needed recreation and their continued existence contributes to stabilizing rural communities.

Forests are managed for timber production and species, habitat and process conservation. A subtle shift from *multiple-use management* to *ecosystems management* is being observed and the new ecological perspective of *multi-functional forest management* is based on the principles of ecosystem diversity, stability and elasticity, and the dynamic equilibrium of primary and secondary production. Making full use of new technology is one of the challenges facing forest management today. Resource information must be obtained with a limited budget. This requires better timing of resource assessment activities and improved use of multiple data sources. Sound ecosystems management, like any other management activity, relies on effective forecasting and operational control.

The aim of the book series ***Managing Forest Ecosystems*** is to present state-of-the-art research results relating to the practice of forest management. Contributions are solicited from prominent authors. Each reference book, monograph or proceedings volume will be focused to deal with a specific context. Typical issues of the series are: resource assessment techniques, evaluating sustainability for even-aged and uneven-aged forests, multi-objective management, predicting forest development, optimizing forest management, biodiversity management and monitoring, risk assessment and economic analysis.

For other titles published in this series, go to
www.springer.com/series/6247

Gherardo Chirici · Susanne Winter
Ronald E. McRoberts
Editors

National Forest Inventories: Contributions to Forest Biodiversity Assessments

EUROPEAN SCIENCE FOUNDATION
SETTING SCIENCE AGENDAS FOR EUROPE

ESF provides the COST Office through an EC contract

COST is supported by the EU RTD Framework programme

Springer

Editors
Gherardo Chirici
Dipartimento di Scienze e Tecnologie
per l'Ambiente e il Territorio
Università degli Studi del Molise
Italy
gherardo.chirici@unimol.it

Ronald E. McRoberts
Forest Inventory and Analysis
Northern Research Station
USDA Forest Service
USA
rmcroberts@fs.fed.us

Susanne Winter
Technische Universität München
Department für Ökologie und
Studienfakultät für Forstwissenschaft und
Ressourcenmanagement
Germany
winter@wzw.tum.de

ISSN 1568-1319 e-ISSN 1568-1319
ISBN 978-94-007-0481-7 e-ISBN 978-94-007-0482-4
DOI 10.1007/978-94-007-0482-4
Springer Dordrecht Heidelberg London New York

Printed on acid-free paper

Springer is part of Springer Science+Business Media (www.springer.com)

Preface

The loss of forest biodiversity stems largely from direct and indirect human activities including deforestation, fragmentation, the degradation of forest habitat and the introduction of invasive species. The current focus is on halting this loss of forest biodiversity and on methods for monitoring the maintenance of forest ecosystems at the international level.

Information required for sustainable forest management as a whole, as well as for forest biodiversity at national and international levels, must be based on robust, statistically sound, updated and long term information systems. National Forest Inventories (NFIs) all contribute to this effort. In many countries inventories have been carried out at national levels and are now increasingly major participants in international reporting mechanisms. Due to different historical backgrounds and varying environmental conditions NFI's use different basic definitions and methods which lead to inconsistencies and lack of comparability for international reporting.

To harmonize global forest information the Food and Agricultural Organization of the United Nations (FAO) has been developing common key definitions. Nevertheless, for practical application for international reporting purposes, these definitions have not been consistently applied. Therefore, the European NFIs decided to collaborate on harmonization of forest information and established the European National Forest Inventory Network (ENFIN).

In addition to its overall mission to provide harmonized forest inventory information on European forests, ENFIN promotes knowledge-sharing, enhanced sampling and assessment methods, and new ideas, thereby maintaining and improving updated forest information systems. It ensures continuous improvement of methods, data collection and data analysis within the NFIs.

Many research projects have developed efficient, optimized methods for monitoring forest biodiversity at a variety of spatial and temporal scales. Nevertheless, a straight forward approach for monitoring forest biodiversity is still lacking. Working Group 3 of COST Action E43 demonstrated that NFI data can be used to achieve comparable and meaningful biodiversity assessments for a large range of selected variables.

This book provides a comprehensive and informative overview of forest biodiversity. In addition, it provides in-depth descriptions and analyses of the essential features of forest biodiversity indicators, the need for harmonized estimates and NFI applications based on practical tests with raw data. Recommendations on the feasibility of forest biodiversity indicators offer a valuable basis for future adoptions of both NFI assessment catalogues and international reporting requirements. I congratulate the editors and authors of this outstanding work on future global monitoring of forest biodiversity.

Vienna
June 2010

Klemens Schadauer
ENFIN Chair

Preface – COST Action E43

The demands for global-level forest information have increased during the past decades due to international agreements and associated reporting requirements. Information reported from different countries should be comparable and on sound statistical bases to be applicable for decision-making.

National Forest Inventories (NFI) have produced forest-related information in some countries for more than 100 years. European NFI teams met in Vienna in 2003 to discuss the new challenges and the measures necessary to promote full use of existing NFIs by data users. As a result, the European National Forest Inventory Network (ENFIN), a network of NFIs, was established. ENFIN members applied for funding to support collaborative efforts to make inventory data and estimates from different countries comparable and inventory results more applicable for users. COST – European Cooperation in Science and Technology – provided the financial means to cover the additional costs needed.

A total of 27 European countries joined COST Action E43, *Harmonisation of National Forest Inventories in Europe: Techniques for Common Reporting*. In addition, the Forest Inventory and Analysis (FIA) programme of the U.S. Forest Service, Scion from New Zealand and the European Joint Research Centre, Institute for Environment and Sustainability, joined COST Action E43 as institutions from non-COST countries. Further, NFI representatives from several other countries participated in the meetings and work of COST Action E43.

COST Action E43 worked closely with international organizations and institutions such as the United Nations Food and Agricultural Organization, the European Commission and the European Environment Agency.

COST Action E43 adopted a mission to develop methods, concepts and definitions for use in harmonizing NFIs so that information from different countries would become fully comparable. The work was organized into three Working Groups: Working Group 1 focused on basic forest inventory concepts and definitions; Working Group 2 focused on forest inventory issues related to greenhouse gas reporting for UNFCCC; and Working Group 3 focused on biodiversity indicators that could be estimated from NFI observations. The work was carried out in meetings, workshops and scientific missions. The official duration of COST Action E43 was from June 2004 to December 2008, but the publishing work continued into 2011.

Members of COST Action E43 collected a large amount of information from the NFIs of the participating countries using questionnaires directed to a wide group of NFI data producers. The information concerned all the main areas of the three Working Groups. Scientific outcomes have been published in multiple journals with a collection of articles published in a special issue of Forest Science. The operational practices of a large number of NFIs were reported as NFI reports in a book published by Springer in 2010[1]. Reports from Brazil, Canada, China, Japan, Luxembourg, Poland, the Republic of Korea and the Russian Federation, in addition to reports for the 29 participating countries and institutions were included.

The three Working Groups worked closely together, particularly in developing common concepts, definitions and methods. However, the work of Working Group 3 was somewhat different from that of Working Groups 1 and 2. Initially, forest biodiversity concepts as they related to forest inventory variables had not been elaborated as much as the basic forest inventory concepts and concepts related to greenhouse gas reporting. One reason is the multidimensionality of biodiversity itself which ranges from genetic diversity to landscape diversity. Working Group 3 focused its efforts at plot-level and landscape diversity by first identifying the most important NFI variables related to forest biodiversity and then deriving seven essential biodiversity features from variables that are relevant for assessing the status of biodiversity. Working Group 3 further assessed the harmonization status of those variables and features among the participating countries, as well as future prospects for harmonized assessments of forest biodiversity using information for variables collected in NFIs. Further, the Working Group demonstrated how harmonized forest biodiversity estimates can be obtained when data from different countries have been collected using different definitions of the basic variables.

This volume is a comprehensive documentation of the work of Working Group 3 of COST Action E43, as well as general information related to forest biodiversity, its assessment and reporting. The participation of forest inventory experts, directly involved in practical work, stimulated the work and promoted successful outcomes. As a chair of the Management Committee of COST Action E43, it is my privilege and pleasure and to thank the editors and authors of this book, as well as the members of Working Group 3 of Cost Action E43, for the outstanding work.

Helsinki
June 2010

Erkki Tomppo

---

[1]Tomppo E, Gschwantner T, Lawrence M, McRoberts RE (eds) (2010) National forest inventories: pathways for common reporting. Springer. 612 p. ISBN 978-90-481-3232-4.

Acknowledgements

This book is the result of the activities of Working Group 3 (WG3): "Contributions of field data acquired in NFI for forest biodiversity assessments" led by Gherardo Chirici and Susanne Winter of COST Action E43 "Harmonisation of National Inventories in Europe: Techniques for Common Reporting" (chaired by Prof. Erkki Tomppo). The on-line interface and the first on-line questionnaire database system were implemented by Alessandro Mastronardi (GM Studio, Italy) who also elaborated the cover picture, thanks to him. We wish to thank all the people who responded to the questionnaires and/or participated in the activities of WG3 of COST Action E43: Albertas Kasperavicius for Lithuania, Anna-Lena Axelsson for Sweden; Annemarie Bastrup-Birk and Vivian Kvist Johannssen for Denmark; Attila Kismarczi for Hungary, Catherine Cluzeau and Nabila Hamza for France; Jacques Rondeux and Christine Sanchez for Belgium; Elmar Hauk for Austria; Iciar Alberdi Asensio, Sonia Condés and Santiago Saura for Spain; Ioizos Ioizou for Cyprus; Jana Beranova for Czech Republic; Jan-Erik Ørnelund Nilsen for Norway; Marko Kovac for Slovenia; Nadia Barsoum for UK; Dieter Pelz for Germany; Tarmo Tolm for Estonia; Teocharis Zagas for Greece; Tiina Tonteri and Helena Makela for Finland; Urs-Beat Brändli for Switzerland; Vladimir Caboun for Slovakia; Paulo Godinho Ferreira for Portugal; and Vittorio Tosi, Patrizia Gasparini, Anna Barbati, Piermaria Corona, Marco Marchetti, Roberta Bertini, Nicola Puletti, and Ugo Chiavetta for Italy.

A special thank you to Nicola Puletti, Anna-Lena Axelsson, Jan-Erik Ørnelund Nilsen, Harald Schmitz and Wolf-Ulrich Kriebitzsch for their contributions to complete the compilation of natural life spans of the dominant tree species by forest category.

We are grateful to Dr. Roberta Bertini who typeset this volume.

All the chapters of this book were peer-reviewed. We would like to thank the following for their revision and useful comments: Prof. Annemarie Bastrup-Birk (University of Copenhagen, Denmark), Dr. Georg Frank (Institut für Waldinventur, Austria), Prof. Federico Maetzke (University of Palermo, Italy), Prof. Roberto Scotti (University of Sassari, Italy), Susan Stein (U.S. Forest Service, USA), Dr. James Westfall (U.S. Forest Service, USA), and Prof. Dr. Barbara Wolff (Applied University of Eberswalde, Germany).

We express our sincere thanks to Catherine Cotton and Ineke Ravesloot at Springer and Selvakumar Gavaskar at SPi Global for their kind assistance in the publishing phase.

June 2010

Gherardo Chirici
Susanne Winter
Ronald E. McRoberts

Contents

Contributors

Meinrad Abegg
Swiss Federal Institute for Forest, Snow and Landscape Reserach (WSL), Schweizerisches Landesforstinventar, Switzerland
meinrad.abegg@wsl.ch

Iciar Alberdi Asensio
Departamento Sistemas y Recursos Forestales, Centro de Investigación Forestal, Instituto Nacional de Investigación y Tecnología Agraria y Alimentaria, Spain
alberdi.iciar@inia.es

Anna Barbati
Università degli Studi della Tuscia, Italy
barbati.sisfor@unitus.it

Nadia Barsoum
Forest Research, Forestry Commission, United Kingdom
nadia.barsoum@forestry.gsi.gov.uk

Annemarie Bastrup-Birk
University of Copenhagen, Faculty of Life Sciences, Forestry and Wood Products, Denmark
ab@life.ku.dk

Jana Beranova
IFER, Institute of Forest Ecosystem Research Ltd, Czech Republic
iana.beranova@ifer.cz

Roberta Bertini
Università degli Studi di Firenze, Italy
bertini.roberta@gmail.com

Urs-Beat Brändli
Swiss Federal Institute for Forest, Snow and Landscape Reserach (WSL), Schweizerisches Landesforstinventar, Switzerland
urs-beat.braendli@wsl.ch

Ugo Chiavetta
Università degli Studi del Molise, Italy
uchiavetta@unimol.it

Gherardo Chirici
Università degli Studi del Molise, Italy
gherardo.chirici@unimol.it

Catherine Cluzeau
IFN, French National Forest Inventory, France
catherine.cluzeau@ifn.fr

Sonia Condés
Escuela Técnica Superior de Ingenieros de Montes,
Universidad Politécnica de Madrid, Spain
sonia.condes@upm.es

Piermaria Corona
Università degli Studi della Tuscia, Italy
piermaria.corona@unitus.it

Nabila Hamza
IFN, French National Forest Inventory, France
nabila.hamza@ifn.fr

Elmar Hauk
BFW, Federal Research and Training Centre for Forests,
Natural Hazards and Landscape, Austria
elmar.hauk@bfw.gv.at

Ronald E. McRoberts
Forest Inventory and Analysis, Northern Research Station,
USDA Forest Service, USA
rmcroberts@fs.fed.us

Marco Marchetti
Università degli Studi del Molise, Italy
marchettimarco@unimol.it

Stefan Neagu
ICAS, Forest Research and Management Institute, Romania
biometrie@icas.ro

Jan-Erik Ørnelund Nilsen
Norwegian Forest and Landscape Institute, Norway
jan-erik.nilsen@skogoglandskap.no

Nicola Puletti
Università degli Studi di Firenze, Italy
nicola.puletti@unifi.it

Jacques Rondeux
Gembloux Agro-Bio Tech, Université de Liège, Belgium
jrondeux@ulg.ac.be

Christine Sanchez
Gembloux Agro-Bio Tech, Université de Liège, Belgium
sanchez.c@fsagx.ac.be

Santiago Saura
Escuela Técnica Superior de Ingenieros de Montes,
Universidad Politécnica de Madrid, Spain
santiago.saura@upm.es

Davide Travaglini
Università degli Studi di Firenze, Italy
davide.travaglini@unifi.it

Susanne Winter
Technische Universität München, Department für Ökologie und Studienfakultät
für Forstwissenschaft und Ressourcenmanagement, Germany
winter@wzw.tum.de

Abbreviations

age_{dom}	Dominant age
age_{stand}	Stand age
age_{tree}	Tree age
AT	Austria
BE	Belgium
CBD	Convention on Biological Diversity
CH	Switzerland
COST	Cooperation in Science and Technology
cwd	Coarse woody debris
CZ	Czech Republic
DB	Data base
dbh	Diameter at breast height
DE	Germany
DK	Denmark
dwd	Harmonized plot-level estimates of deadwood volume per unit area
EE	Estonia
EEA	European Environmental Agency
EFT	European Forest Types (sensu EEA 2006)
ENFIN	European National Forest Inventory Network
ES	Spain
EU	European Union (in some Figures and Tables for EU we intended the pan-European area, this is clarified in the text)
FAO	Food and Agriculture Organization of the United Nations
FHH	Fauna-Flora-Habitat
FI	Finland
FIA	Forest Inventory and Analysis programme, U.S. Forest Service
FR	France
FRA	Forest Resource Assessment
GR	Greece
GPG	Good Practice Guidance
HU	Hungary

ICP Forests	International Co-operative Programme on Assessment and Monitoring of Air Pollution Effects on Forests operating under the UNECE Convention on Long-range Transboundary Air Pollution
IPCC	Intergovernmental Panel on Climate Change
IT	Italy
LIS	Line Intersect Sample
LT	Lithuania
LULUCF	Land Use, Land Use Change and Forestry
LV	Latvia
MCPFE	Ministerial Conference on the Protection of Forest in Europe
NFI	National Forest Inventory
NO	Norway
PT	Portugal
Rf	Reducing factor
RO	Romania
SEBI 2010	Streamlining European 2010 Biodiversity Indicators
SE	Sweden
SFM	Sustainable Forest Management
SI	Slovenia
SK	Slovakia
TBFRA 2000	Temperate and Boreal Forest Resources Assessment 2000
UNECE	United Nations Economic Commission for Europe
UNEP	United Nations Environment Programme
UK	United Kingdom of Great Britain and Northern Ireland
USA	United States of America
WG	Working Group

Chapter 1
The Need for Harmonized Estimates of Forest Biodiversity Indicators*

Susanne Winter, Ronald E. McRoberts, Gherardo Chirici, Annemarie Bastrup-Birk, Jacques Rondeux, Urs-Beat Brändli, Jan-Erik Ørnelund Nilsen, and Marco Marchetti

Abstract Forest biodiversity is crucial to the ecological, economic, and social well-being of earth's civilisations. Unfortunately, however, forest biodiversity is threatened to a serious degree in nearly all countries. Therefore, many countries have agreed to be parties to international agreements focused on maintaining, restoring, and monitoring biodiversity; further, these countries have agreed to report to international bodies on the status and trends in forest biodiversity. NFIs are the primary source of large-scale information available for this purpose, but the large variety of definitions, protocols, sampling designs, and plot configurations used by

* This publication is supported by COST.

S. Winter (✉)
Department für Ökologie und Studienfakultät für Forstwissenschaft und Ressourcenmanagement, Technische Universität München, Germany
e-mail: winter@wzw.tum.de

R.E. McRoberts
Forest Inventory and Analysis, Northern Research Station, USDA Forest Service, USA
e-mail: rmcroberts@fs.fed.us

G. Chirici and M. Marchetti
Università degli Studi del Molise, Italy
e-mail: gherardo.chirici@unimol.it; marchettimarco@unimol.it

A. Bastrup-Birk
University of Copenhagen, Faculty of Life Sciences, Forestry and Wood Products, Denmark
e-mail: ab@life.ku.dk

J. Rondeux
Gembloux Agro-Bio Tech, Université de Liège, Belgium
e-mail: jrondeux@ulg.ac.bc

U.-B. Brändli
Swiss Federal Institute for Forest, Snow and Landscape Reserach (WSL), Schweizerisches Landesforstinventar, Switzerland
e-mail: urs-beat.braendli@wsl.ch

J.-E.Ø. Nilsen
Norwegian Forest and Landscape Institute, Norway
e-mail: jan-erik.nilsen@skogoglandskap.no

G. Chirici et al. (eds.), *National Forest Inventories: Contributions to Forest Biodiversity Assessments*, Managing Forest Ecosystems 20,
DOI 10.1007/978-94-007-0482-4_1,

NFIs makes comparable international reporting extremely difficult. COST Action E43 was initiated to address this problem by developing harmonization techniques that facilitate common reporting. Harmonization typically consists of two components: development of common international reference definitions and development of bridging techniques that facilitate estimation according to reference definitions using data collected according to national definitions. Working Group 3 of COST Action E43 has focused its harmonization efforts on issues related to biodiversity. The chapters and sections that follow document these efforts in detail.

1.1 Forest Biodiversity

1.1.1 Introduction

Forest ecosystems are among the most biologically rich and genetically diverse terrestrial ecosystems on earth. Of the 38 main classes in Holdridge's (1947, 1967) life zone classification, more than half (19 forest and two woodland formations) are dominated by trees. The World Wildlife Fund (Dinerstein et al. 1995) identified 14 major earth habitat types of which seven are forest types. Depending on definitions, 22–30% of the earth's surface is covered by forests and wooded lands (FAO 2005; GFW 2006), and these lands provide habitat for 70% of known animal and plant species (Matthews et al. 2000). Thus, forests make substantial contributions to global biodiversity.

However, forest diversity is increasingly threatened with at least one tree species at risk in each country of the world (FAO 2005). South and Southeast Asia have the greatest absolute number of threatened tree species (403), whereas the greatest percentages of tree species threatened by extinction are in the Caribbean region (15%) and in North and Central America (12%). A major driver for species loss worldwide is habitat degradation and loss (Foley et al. 2005). Earth's tropical forest was diminished by 50% between 1978 and 1988 (Skole and Tucker 1993), and, on average, 2.3 million ha of tropical forest were lost annually between 1990 and 1997 because of fragmentation, burning, logging, and conversion to other uses (Achard et al. 2002). Thus, conservation and enhancement of forest biodiversity is crucial to maintaining forest health and the global ecological balance, sustaining the production of raw materials for forest-based industries and providing other goods and services. At the European level, the intent to track progress towards halting the loss of biodiversity by 2010 (UNECE 2003) clearly expresses a crucial objective of global civilisation (Balmford et al. 2005; Mikusinski et al. 2007).

1.1.2 What is Forest Biodiversity?

One broad and recognised definition of forest biodiversity refers to "the diversity of life in all its forms and all its levels of organization within forested areas" (Hunter 1990). This definition means that forest biodiversity encompasses trees and other plants, animals, and micro-organisms that inhabit forest areas. Further, biodiversity may be defined using multiple independent sets of theoretical concepts. Three major definitional approaches refer to the scale and the extent of forest biodiversity. Whittaker (1972) defines biodiversity using three spatial types: *alpha* (α) diversity, which refers to ecosystem diversity; *beta* (β) diversity, which refers to the change in diversity between ecosystems; and *gamma* (γ) diversity, which refers to the overall diversity for different ecosystems within a region. Noss (1990) defines diversity in terms of three classes or components: *compositional*, which relates to the identity and variety of elements; *functional*, which relates to ecological and evolutionary processes; and *structural*, which relates to the physical organization of the pattern of elements. Finally, Leveque (1994) and Gaston and Spicer (2004) use a systematic approach that characterizes three levels of biodiversity: *genetic*, *species* (or taxonomic) and *ecosystem*. The Whittaker (1972) approach is the most common and least complex approach, although the Noss approach (1990) has been widely used to describe the structural components of forest biodiversity using existing measurements.

Any definition of forest biodiversity first requires a definition of *forest*, and forest, in turn, requires a definition of *tree*. For this book, the focus is on definitions of forest, with the definition of tree referred to Gschwantner et al. (2009) to avoid an excessive cascade of definitions. Although definitions of forest are numerous (Kleinn 1991; Lund 2006) the most commonly used criteria include tree density which is primarily assessed using crown cover or stem densities, area, and the average natural height of trees (Vidal et al. 2008). However, definitions of forest vary with respect to study purposes and scales (Mathys et al. 2006), an inconsistency that hampers surveys of forest biodiversity at large scales (Puumalainen et al. 2003). For future reference, we use the COST Action E43 definition (Vidal et al. 2008):

> *Forest is land spanning more than 0.5 ha with trees higher than 5 metres and with tree crown cover of at least 10%, or able to satisfy these thresholds in situ. For tree rows or shelterbelts, a minimum width of 20 m is required. Forest does not include land that is predominantly agricultural or urban land use.*

1.1.3 Why maintain Forest Biodiversity?

Hunter (1999) provides multiple reasons for maintaining forest biodiversity of which the primary is economic value. From a human perspective, forest biodiversity is the basis for medicines; for industrial material; for food and drink; for spiritual value including leisure, culture and aesthetics; for scientific and educational values such as bionics; and for ecological value.

Considering only the plant species component of biodiversity, approximately 415,000 species of vascular plants (spermatophyta and pteridophyta), bryophytes, lichens, fungi, algae, and flagellates are recorded on earth (BfN 2002) with numbers generally increasing from the poles to the equator. Europe and the United States of America (USA) provide forest habitat for 200–3,000 vascular plant species per km^2 (Barthlott et al. 1999). Flora and fauna, forest structure and all functional processes starting from the decomposition of geological material, soil development, and climatic and biochemical reactions are components of forest biodiversity. If halting the loss of biodiversity is a scientific, social, or economic human objective as required by UNECE (2003), then forest biodiversity must be monitored, maintained, and if necessary restored.

1.2 The Role of National Forest Inventories in Monitoring Biodiversity

1.2.1 Biodiversity and National Forest Inventories

Traditionally, wood production has been the most important forest function in much of the world. With industrialization, the increasing demand for wood led to the organization and legal regulation of forest management which, in turn, led to the first forest inventories. Historically, forest inventories were conducted to obtain information necessary for management at the local level. Large scale inventories date back only to the early twentieth century: 1919 in Norway, 1921 in Finland, 1923 in Sweden (Köhl and Brassel 1999), and 1928 in the USA. During the later twentieth century, many European countries developed and implemented national forest inventories (NFIs) by applying the sampling and methodological lessons learned from local forest management inventories. However, information from local management inventories cannot be readily aggregated to the national level to produce defensible estimates comparable to those obtained from NFIs (Adermann 2010; Tomppo et al. 2001).

Carlowitz (1713) first formulated the principle of sustainable forest management (SFM) as a means of guaranteeing a continuous wood supply. The basis for SFM is information on the state and changes in timber stock, increment and yield. Incorporation of objectives related to biodiversity, which is now a component of SFM, occurred even before the term biodiversity was first used (Franklin 1988). One primary biodiversity objective is to integrate biodiversity monitoring and assessment into forest management strategies and activities (Puumalainen et al. 2003; Winter et al. 2004). The transition from traditional forest management and inventorying for wood production purposes to multi-purpose resource monitoring that incorporates SFM, including biodiversity objectives, has required the development of new NFI objectives, new sampling designs, and new estimation procedures (Köhl et al. 1995; Iles 1998; Lund 1986). Because these redesigned NFIs are more comprehensive than wood production inventories, they are increasingly regarded as the primary source of reliable data for guiding SFM.

Development of a suitable method for monitoring biodiversity has been a topic of concern since the term biodiversity was first used (Chadwick et al. 1999). The forest biodiversity concept incorporates aspects of structural diversity such as species richness, vertical and horizontal architecture, physical status (standing/lying, living/dead/decomposing), and aspects of functional diversity such as genetic diversity and ecological and evolutionary processes (Wilson 1992; Wilson et al. 1996). Regardless of the variety of forest biodiversity components, biodiversity must be monitored in strict compliance with accepted concepts to facilitate reporting, to evaluate current status and trends, and to select practical maintenance measures. Relative to Noss' (1990) hierarchical biodiversity concept, recent proposals for acquiring biodiversity assessment data are bottom up in the sense that they scale local observations to landscape levels. Under this concept, diversity is monitored using an approach based on a minimum core set of tasks and attributes but which permits inclusion of additional objectives (Green et al. 2005; Teder et al. 2007). However, large scale, quantifiable objectives for monitoring forest biodiversity that provide unambiguous measures of progress are still lacking (Lindenmayer et al. 2008).

1.2.2 Forest Biodiversity Indicators

Acquisition of data by NFIs to assess all aspects of forest diversity is not feasible (Rondeux 1999); following Boutin et al. (2009): "it will never be possible to measure all that is biodiversity". Thus, the emphasis turns to acquiring data to estimate indicators that aggregate and synthesize information for multiple aspects of forest biodiversity simultaneously. An indicator is a quantitative or qualitative variable that can be measured or described and which, when observed periodically, demonstrates trends (Montréal Process 2005). Ideally, indicators should be appropriate for local scales but should also provide information that can be aggregated for larger scales. In addition, to ensure that future assessments are cost efficient and practicable for obtaining time series of data to estimate trends, indicators should be based on components of forest biodiversity that can be readily estimated using data collected by standard forest inventories. Using NFI data to estimate forest biodiversity indicators may permit comparisons in space between local, national and international regions and also in time. Thus, the challenge is to develop forest biodiversity indicators and monitoring methods that facilitate comparisons in both space and time (Brändli et al. 2007; Newton and Kapos 2002). Many indicators have been proposed (e.g. Bosch and Söderbäck 1997; UNEP 2001; MCPFE 2003; Montréal Process 2005), and an overview of recommended indicators has been provided by the European Environment Agency (EEA 2003, 2009).

1.2.3 Geographic Scale

Geographic scale is a key aspect of forest biodiversity monitoring, because interactions and processes underlying biodiversity vary according to whether the scale is national (10s of millions km^2), regional (1,000 ha to millions km^2), landscape (100–1,000 ha) or patch (1–100 ha) (Williams 2004). A consequence is that forest management methods selected to achieve biodiversity objectives should be matched to scale (Lindenmayer et al. 2008). For example, a European beech (*Fagus sylvatica*) stand in the lowlands of central Europe may also have a small proportion of other species such as Sessile oak (*Quercus petrea*), European hornbeam (*Carpinus betulus*) and Sycamore maple (*Acer pseudoplatanus*). Gaps could be cut in the stand to enrich biodiversity, and alien or exotic tree species such as Douglas fir (*Pseudotsuga menziesii*) and Japanese larch (*Larix kaempferi*) could be planted in the gaps of the natural plant community. At local spatial scales, tree biodiversity would increase. However, at larger spatial scales, animal species requiring large natural habitats will lose habitat and specialized species will find their habitat degraded; both cases have negative consequences for population vitality. Thus, local biodiversity enhancement often results in regional biodiversity degradation. Similarly, management decisions to increase large-scale biodiversity could be desirable and sensible but at the local scale, the same decision might be completely inappropriate.

Different challenges and different objectives characterize biodiversity monitoring and conservation at different scales. First, at the global scale, monitoring has been incomplete. The number of genes, taxons and species, functions and structures is unknown (Gaston and Spicer 2004). Reports at the global level are primarily based on national data collected for broad overviews (e.g., FAO 2005) or narrowly focused assessments of the biodiversity status and threats for single species groups such as amphibians (Gallant et al. 2007) or higher plants (BfN 2002). No consensus exists regarding a global monitoring approach.

Second, biodiversity monitoring at the continental scale must focus on the specifics of each continent. For example, European forests were intensively impacted during the last glacial periods, and as a consequence, fewer tree and other species are found in Europe than in northern Asia or North America. In Europe, only a few tree species were sufficiently competitive to spread over large regions following the glacial period; of these, European beech (*Fagus sylvatica*) is found naturally throughout central Europe without a real competitor. Because of the uniqueness of this phenomenon, beech forests are the focus of the European Fauna-Flora-Habitat (FHH) directive (Council of the European Union 1992) and e.g. German law (BNatSchG 2002, §32) that requires conservation of all remnants of old beech forests. Forest biodiversity monitoring at the continental level requires concepts that accommodate all biogeographical regions. In Europe, a suitable monitoring approach must accommodate forests that range from macaronesian to boreal with their associated differences in tree and herb species composition, forest structure and life cycles.

Third, at the biogeographic scale, the FHH directive divides Europe into seven regions with similar growth conditions within regions. All efforts to conserve European forest biodiversity are focused on these regions to ensure sensible international cooperation. Fourth, at the national scale, international commitments must be adopted by individual countries which are then responsible for implementation. However, national policies often supersede international decisions to which countries are parties. Fifth, at the local scale, preserving and monitoring biodiversity is foundational to maintaining earth's biodiversity. Good management, which includes successful monitoring of the selected measures, requires precise understanding of biodiversity requirements, commitments and laws at the other levels.

1.2.4 The Challenges

A non-trivial and crucial issue for designing a successful biodiversity monitoring procedure is the selection of the biodiversity objectives for which data are to be collected (Kovac et al. 2007). Forest managers have different requirements than forest ecologists who may be primarily interested in forest ecosystem composition, functions, and their dynamics. In addition, whereas managers and some scientists often select variables to assess individual aspects of biodiversity, others favour a holistic approach and select variables that integrate many components and give insight into the working of whole systems (Kovac et al. 2007; Corona and Marchetti 1998). The solution is to design and implement biodiversity monitoring methods that are appropriate for multiple land uses, multiple objectives, multiple scales, and multiple assessment approaches.

A second crucial aspect of designing a successful biodiversity monitoring program is development of appropriate sampling strategies. Important components of sampling strategies include distribution of plots across the landscape, selection of a sampling method (e.g., fixed area plots, transects, Bitterlich sampling), selection of appropriate plot configurations including size and shape, and measurement protocols. For example, larger plots may be necessary to assess forest biodiversity components such as tree species composition, horizontal and vertical structural diversity and forest age, whereas smaller plots may be more efficient for assessing components such as regeneration, ground vegetation including lichens and fungi, and microhabitats. In addition, whereas plots for assessing the previously noted components are usually configured as circles, line intersect sampling is also common for assessing components such as lying deadwood. Finally, attributes such as naturalness that integrate multiple components of forest biodiversity require information that is collected from plots of multiple sizes and configurations. Thus, it is clear that any adequate and proper sampling design that produces the information necessary to satisfy forest biodiversity reporting requirements will, of necessity, be complex.

Sampling designs that are optimized to accommodate individual biodiversity components would be vastly different, an outcome that time and financial resource constraints simply do not permit. Therefore, compromises that accommodate both the need for data that produce accurate estimates for biodiversity indicators and specified time and cost constraints are necessary. An overview of the variety of sampling designs used by NFIs for these purposes is provided in Chap. 3, whereas specific features of sampling designs for individual countries are described in Tomppo et al. (2010: Chap. 2, NFI Reports section). Nevertheless, some features of biodiversity sampling strategies perhaps should not be compromised. For example, many well-accepted biodiversity indicators are expressed in terms of trends which require multiple measurements of the same sampling units over time. Thus, one component of a successful biodiversity monitoring program would be inclusion of at least a proportion of permanent plots. As a second example, remotely sensed data, often satellite imagery obtained from sensors such as Landsat and SPOT, have been found to be particularly useful for assessing landscape features such as fragmentation and patch size (Meneguzzo and Hansen 2009; Nelson et al. 2009). Selection of ground sampling strategies that support and facilitate remote sensing should also be considered (Nilsson et al. 2003). However, an important point is that while accommodating measurement devices such as remote sensors, assessments should be independent of the devices (Tomppo et al. 2010). Development of sampling strategies for biodiversity monitoring will include many challenges.

NFIs are a rich and comprehensive source of forest biodiversity data. They already provide information on the key components of forest ecosystems such as the volume of tree growing stock by species, forest type and forest structure, ground vegetation, soil, and site conditions (Rondeux 1999). In addition, they provide information on spatial arrangements of ecosystems, tree species composition, and landscape scale elements (O'Neill et al. 1988). Finally, NFIs already collect information for many of the variables that are most important for assessing biodiversity (Chap. 2, this book; Winter et al. 2008). From an efficiency perspective, NFIs may be considered the starting point for investigations of forest biodiversity.

1.3 International Reporting Requirements

1.3.1 International Agreements

Although historically timber production has been regarded as the primary function of forests, recent years have seen a shift to a more multi-functional and holistic view of forest resources. With this view, ecosystem services such as recreation, health and well-being, protection against environmental risks, biological diversity, and mitigation of climate change effects are increasingly recognized as integral components of sustainable forest management (UNEP FI 2007). This role of forest management and planning in conserving forest biodiversity is reflected in commitments by many governments to biodiversity and SFM initiatives: (1) the Ministerial Conference on the Protection of Forests in Europe (MCPFE 2002) which includes, as a criterion, *Maintenance, conservation, and enhancement of biological diversity in forested ecosystems*; (2) the Montréal Process (2006) which covers temperate and boreal forests and includes the criterion *Conservation of biological diversity*; (3) the Convention on Biological Diversity (CBD 1992) signed by 198 countries and the European Union; (4) Natura 2000, a European ecological network of special areas of conservation (Heath et al. 2000; Ssymank et al. 1998) and natural habitats monitored to ensure maintenance or restoration of their composition, structure and extent (EEC 1992); and (5) the FFH directive which defines the need for the conservation of habitats and species with the adoption of appropriate measures (Council of the European Union 1992).

Many of the international agreements that embody these commitments and initiatives require periodic reports of estimates of national forest resources, particularly as they relate to SFM and biodiversity. The primary SFM conventions, the Montréal Process (2005) and the Ministerial Convention on Protection of Forests in Europe (MCPFE 2003), require that member countries report on SFM and biodiversity indicators. The Convention on Biological Diversity (CBD 2007) requires that countries identify and monitor components of biological diversity for purposes of conservation and sustainable use. The Streamlining European 2010 Biodiversity Indicators (SEBI 2010) initiative is the means by which the European Commission, through the European Environmental Agency (EEA), is implementing the Strategic Plan for the Conservation of Biological Diversity of the CBD. SEBI (2010) developed a set of 26 indicators to track temporal biodiversity changes as a means of monitoring progress toward achieving the 2010 European objective of halting the loss of biodiversity (EEA 2007, 2009). Although many of these indicators are general, two specifically relate to forests: (1) forest growing stock volume, increment and fellings, and (2) forest deadwood. Parties to the United Nations Framework Convention on Climate Change (UNFCCC) (1992) are required to produce annual reports of greenhouse gas emissions and removals by sources and sinks. Finally, following the CBD (1992) and Helsinki Resolution H1 (1993) on SFM, most European countries must now report on numerous indicators of forest health, biodiversity, and functions such as recreation, soil and water protection, and wood production.

NFIs are the sources of the most extensive and comprehensive data on the status of forests in Europe, North America and other regions for reporting under international agreements. Although NFIs share a primary objective of conducting forest resource assessments to describe forest structure and to assess SFM, they do not assess common sets of variables or use common sampling designs, plot configurations, measurement protocols, or analytical methods. These disparities contribute to the lack of comparability among data and estimates available for international reporting. For example, the 2000 Forest Resource Assessment (FRA) report indicated severe problems in the harmonization of definitions for natural forest, other wooded land and forest available for wood supply and forest area by protection categories. Gabler et al. (submitted) discuss the inconsistency among countries relative to the FAO reference definitions when reporting to the 2005 Global FRA.

1.3.2 The Effects of Disparate NFI Definitions and Methods

Two sensitivity analyses have been conducted to investigate the effects on forest estimates of using different national definitions. Traub et al. (1997) simulated forest cover patterns as a means of investigating the effects of different national forest area definitions on forest area estimates. In general, estimates for scattered, fragmented forests with gradual transitions between forest and non-forest land cover were most sensitive to different definitions. At national levels, the estimate of total forest area for Spain decreased by 8% when the United Kingdom's definition of forest land was used and increased by 6% when Luxembourg's definition was used. At the regional level, the pan-European estimate of forest area decreased by 6% when the United Kingdom's definition was used and increased by 3% when Luxembourg's definition was used. In addition, the national definitions of forest land used by Finland, Sweden, and Norway cannot be applied elsewhere because they include a criterion related to wood production capability and, as such, are not compatible with the definitions of other countries.

Cienciala et al. (2008) used data from the 9th (1996–2003) Finnish NFI to compare carbon pool change estimates based on Finnish national definitions to estimates based on the international definitions used for the 2000 Temporal and Boreal Forest Resource Assessment (TBFRA 2000) (UNECE and FAO 2000). Finland's forest area estimates were 10.6% less when using the Finnish production-oriented definition than when using the TBFRA 2000 definition. Similarly, Finland's estimate of the volume of growing stock volume decreased by 2.7%, and the corresponding annual volume increment decreased by 1.8%. The smaller percentage decreases in volume and volume increment relative to the percentage decrease in forest area are attributed to the inclusion of poorer sites with less mean volume and less mean volume increment under the TBFRA 2000 definition; these sites are not considered forest land when using the Finnish definition.

Two approaches have been proposed for circumventing the effects of different national definitions in the context of international reporting. Köhl et al. (2000) describe the first, *standardization*, as a top-down approach that follows a common system of nomenclature and focuses on common standards with respect to NFI definitions and methods. The second approach, *harmonization*, is based on the acknowledgement that individual countries have developed the unique features of their NFIs for specific purposes and are justified in their desire to maintain them. Thus, the harmonization approach focuses on developing methods for producing comparable estimates despite the lack of standardization. Köhl et al. (2000) describe harmonization as a bottom-up approach that begins in divergence and ends in comparability.

The issue of harmonization has received increased attention in recent years. The United Nations Economic Commission for Europe (UNECE)/FAO team responsible for the TBFRA 2000 compiled international definitions and then harmonized national estimates to conform to them (UNECE and FAO 2000). The Global FRA 2005 (FAO 2005) continued this effort. The Intergovernmental Panel on Climate Change (IPCC) developed monitoring guidelines that address issues of transparent and harmonized reporting. The Good Practice Guidance (GPG) (IPCC 2003) for Land Use, Land Use Change and Forestry (LULUCF) provides detailed guidance for some aspects of reporting. However, despite the efforts of these international organisations to promote and facilitate harmonized reporting, none of these organisations have the resources necessary to construct common definitions or to develop methods that can be implemented on an operational basis (Tomppo et al. 2010).

1.4 COST Action E43

1.4.1 The European National Forest Inventory Network

In response to the requirements for harmonized European reporting, representatives of European NFIs established the European National Forest Inventory Network (ENFIN) in Vienna, Austria, in 2003. The overall objective of ENFIN is to promote NFIs as comprehensive monitoring systems whose forest ecosystem information can be used to address a broad array of forest related issues. ENFIN has four specific aims. First, ENFIN aims at enhancing cooperation among European NFIs as a means of strengthening their capacities to satisfy national, European, and other international requirements for timely and harmonized forest information. Second, ENFIN aims to promote knowledge-sharing and new ideas, thereby ensuring continuous improvement of methods, data collection and data analysis within the NFIs. Third, ENFIN aims at maximizing the synergy between NFIs and other European data collection systems and monitoring and reporting activities. Fourth, ENFIN aims at ensuring openness to new requirements on forest data for emerging policy needs.

ENFIN representatives signed a memorandum of understanding to promote closer long-term collaboration and regular meetings for common discussion and information exchange among European NFIs. ENFIN also agreed to solicit financial support from the European Commission and to initiate an Action under the auspices of the European program *Cooperation in Science and Technology* (COST) (2009). Founded in 1971, COST is an intergovernmental framework to facilitate coordination of European research. COST activities are based on networks of coordinated research projects characterized as Actions that are of interest to member states. COST Actions focus on maximizing synergy, adding value via cooperative research, and promoting integration. Currently, there are more than 200 Actions of which more than 30 deal with forestry issues. Interested institutions from non-COST member countries are welcomed without regard to geographic location.

1.4.2 Background on COST Action E43

In the spring of 2004, ENFIN submitted a proposal for a COST Action, titled *Harmonization of the National Inventories in Europe: Techniques for Common Reporting*. The primary objective of the proposal was to solicit financial support for work on harmonized reporting by NFIs. The proposal was successful, and the first meeting of the Management Committee of the new COST Action E43 was held in Brussels, Belgium, in April 2004.

COST Action E43 has been the most comprehensive effort yet directed toward harmonization of NFIs. The primary objectives of COST Action E43 are threefold: (1) to harmonize existing European NFIs, (2) to support new forest inventories for the purposes of satisfying requirements for providing current and harmonized forest

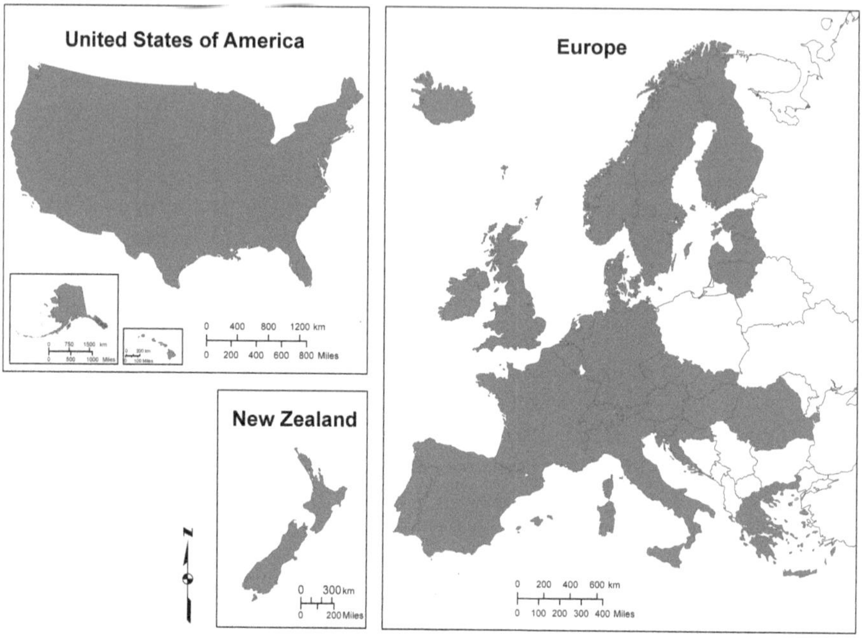

Fig. 1.1 Countries participating in COST Action E43. Data source: ESRI map 2008. Created by Lisa G. Mahal, University of Nevada, Las Vegas (USA)

resource information, and (3) to promote scientifically sound forest inventory designs, data collection, and data analyses (COST E43 2008). Participating institutions included government agencies or universities representing the NFIs of most European countries (Fig. 1.1). Also participating as non-COST institutional members were the Forest Inventory and Analysis (FIA) programme of the U.S. Forest Service; Scion, the New Zealand Forest Research Institute; and the Institute for Environment and Sustainability, Joint Research Centre of the European Commission.

COST Action E43 was organized into three working groups: Working Group 1 (WG1) addressed harmonization of NFI definitions and measuring practices; Working Group 2 (WG2) addressed harmonization of estimation procedures for carbon pools and carbon pool changes using NFI data; and Working Group 3 (WG3) addressed harmonization of indicators and estimation procedures for assessing components of biodiversity using NFI data. A general overview of COST Action E43, its activities, and its accomplishments are found in Tomppo et al. (2010). The sections and chapters that follow focus on the activities of WG3 of COST Action E43.

1.4.3 Working Group 3 of COST Action E43

The primary task of WG3 was to identify possibilities for using NFI field data to produce estimates of biodiversity that are comparable over vegetation zones.

Three primary outputs were expected from WG3: (1) a synthesis of current definitions and practices used to assess biodiversity in different countries, (2) recommendations regarding harmonization techniques that may be used to produce comparable reporting among countries, and (3) a strategy for monitoring and reporting biodiversity. WG3 built on the work, knowledge, and suggestions from multiple other initiatives such as Cost Action E27 (Protected Forest Areas in Europe – Analysis and Harmonization), International Cooperative Program – Forest (ICP-Forests), and national developments that contribute to satisfying the reporting requirements of the TBFRA, MCPFE, and the Montréal Process.

WG3 pursued its objectives via a combination of working group meetings, task force meetings, short-term scientific missions, and voluntary work at home institutions (Table 1.1). The methods used by WG3 to investigate harmonization are briefly discussed in Sect. 1.5, and specific results are reported in subsequent chapters.

Table 1.1 Timeline for working group 3 activities[a]

When	What	Where	Milestones
2004 September 9–10	1st joint working group and management committee meeting	Hørsholm, Denmark	Kick off meeting
2004 October 21–22	2nd joint working group and management committee meeting	Gembloux, Belgium	Discussion of general approach for WG3
2004 December 2–3	3rd joint working group and management committee meeting	Florence, Italy	Development of first questionnaire
2005 April 14–15	4th joint working group and management committee meeting	Vienna, Austria	Selection of essential features and sub-working group leadership
2005 September 19–20	5th joint working group and management committee meeting	Freiburg, Germany	Development of second questionnaire
2006 May 11–13	6th joint working group and management committee meeting	Bordeaux, France	Collection of local definitions of essential features
2006 November 16–18	7th joint working group and management committee meeting	Thessaloniki, Greece	Collection of local definitions of essential features
2006 December 13–14	WG3 task force meeting	Birmensdorf, Switzerland	Discussion of local definitions and possible harmonization strategies
2007 February 22–23	WG3 task force meeting	Vienna, Austria	WG3 rules for managing cross-WG references

(continued)

Table 1.1 (continued)

When	What	Where	Milestones
2007 March 22–24	Joint workshop on assessment of biodiversity in forests in Europe (ICP forests expert panel on biodiversity and ground vegetation/COST action E43-WG3) WG3 task force meeting	Florence, Italy	Cooperation between COST Action E43 and ICP Forest; Development of the common NFI DB structure
2007 June 7–9	8th joint working group and management committee meeting	Haikko, Finland	Presentation of the data acquired and structured in the common NFI DB
2007 30 September-21 October	Short-term scientific mission	Freising, Germany	Construction of the common database
2008 October 25–27	9th joint working group and management committee meeting	Bucharest, Romania	Development of reference definitions and bridges
2008 January 28–29	Joint WG task force and steering committee meeting	Birmensdorf, Switzerland	First drafts of reference definitions for the forest biodiversity indicators. Work on cross-WG references
2008 March 17–19	Editorial board meeting	Helsinki, Finland	Organisation of scientific publications
2008 June 5–7	10th joint working group and management committee meeting	Lisbon, Portugal	Tentative development of bridges for application to common NFI DB; Revision of references
2008 June 1–23	Short term scientific mission	Freising, Germany	Construction of the common database
2008 September 17–19	WG3 task force meeting	Hørsholm, Denmark	First results of test of bridges; Revision of references
2008 October 26–28	Editorial board meeting	Edinburgh, Scotland	Organisation of scientific publications
2009 February 16–18	Editorial board meeting	Freising, Germany	Structure of book

[a] Agendas and minutes of meetings are available at: http://www.metla.fi/eu/cost/e43

1.5 The Harmonization Process

1.5.1 Overview

WG3 conducted its harmonization investigations in four phases. First, WG3 developed a questionnaire that was distributed to the NFIs of participating countries for the purpose of evaluating the importance and feasibility of a large number of variables that are potentially useful for biodiversity assessments. The variables deemed most important and feasible were aggregated into categories corresponding to the components of biodiversity, and common definitions for each variable were developed. Second, the responses to a second questionnaire were used to evaluate agreement among NFIs on these common definitions and measurement practices. Third, bridges (Ståhl et al. submitted) were developed to convert estimates based on national definitions to estimates based on the common definitions. Fourth, and finally, a database of NFI data was constructed and used to test the bridges.

1.5.2 Reference Definitions

Harmonization typically begins with development of common definitions called *reference definitions*. In the development of a reference definition, Vidal et al. (2008) identify nine desirable attributes:

1. *acceptability*, meaning adoption at national and international levels for international reporting;
2. *objectivity*, meaning free of particular interests of individual NFIs or stakeholders;
3. *clearness*, meaning easily grasped and clearly stated;
4. *sufficiency*, meaning covering all relevant cases;
5. *usefulness*, meaning satisfaction of forest, industry, and environmental needs at national and European levels;
6. *sustainability*, meaning long-term validity;
7. *neutrality*, meaning not to be used as a means of assessing the quality of national NFIs or national definitions;
8. *practicality*, meaning NFIs must be able to provide results conforming to the reference definitions; and
9. *independence*, meaning validity is independent of the measurement protocols and instruments.

The process of developing reference definitions begins with questionnaires that inquire regarding aspects of national definitions such as the variables and variable thresholds. The process consists of four steps: (1) review national and international definitions, (2) decompose definitions by listing variables used and create classes for each variable, (3) select relevant variables and thresholds, and (4) construct and revise the reference definition until final acceptance by all countries.

WG3 accepted basic reference definitions developed by WG1 such as those for tree, forest, and growing stock (Vidal et al. 2008; Gschwantner et al. 2009) and focused on developing reference definitions unique to biodiversity assessments. Harmonization investigations by WG3 were conducted in the framework of four reference levels: concept (forest biodiversity), essential feature (e.g., forest structure), indicator (e.g. species composition), NFI variable (tree species) (Table 1.2). Reference definitions at level 4, NFI variables, were accepted as those developed by WG1 and WG2.

In Phase 1, WG3 selected 41 variables potentially relevant for biodiversity assessments on the basis of participant suggestions, literature on biodiversity indicators, and current ecological knowledge. A questionnaire on the 41 variables was developed and distributed to forest inventory experts from NFIs represented among COST Action E43 participants. The questionnaire had three primary objectives: (1) to select important and feasible variables for forest biodiversity assessments currently assessed by participating NFIs, (2) to evaluate the importance of the variables for forest biodiversity assessments, and (3) to determine the primary differences among participating NFIs with respect to the numbers and types of forest biodiversity variables assessed. Responses were received from the NFIs of 25 European countries and the USA. Of the 17 variables that were evaluated as most important and feasible for assessment by NFIs, 13 were assigned to seven categories characterized as essential features of forest biodiversity: forest categories, forest structure, forest age, deadwood, regeneration, ground vegetation and naturalness (Sect. 2.3). Greater detail on the questionnaire and the evaluation of the responses is provided in Chap. 2.

In Phase 2, responses from 25 European countries and the USA to a second questionnaire were used to acquire detailed information concerning the degree to which the 13 important and feasible variables were already harmonized.

Table 1.2 Working group 3 reference level framework for the forest structure essential feature

Reference level			
1 Concept	2 Essential features	3[a] Indicators	4[a] NFI variables
Forest biodiversity	Forest category	–	–
	Forest age	–	–
	Forest structure	Species composition: proportions of species	Tree species
		Vertical structure: number of layers	Tree height
		Horizontal structure: standard deviation of dbh	Tree dbh
	Deadwood	–	–
	Regeneration	–	–
	Ground vegetation	–	–
	Naturalness	–	–

[a] Indicators and NFI variables for other essential features are provided in Chaps. 3 and 5

This questionnaire focused on the methods and thresholds used to observe or measure the 13 variables, the level of expertise necessary, and the combinations of plot components and land use categories for which observations or measurements are obtained. The responses to the questionnaire indicated that most of the 13 variables are already assessed by the NFIs of a large proportion of countries and that measurement techniques were generally similar. However, there was considerable lack of agreement regarding the expertise necessary to assess the variables with eight NFIs indicating the necessity of a high level, most indicating a mid-level of expertise, and two NFIs indicating that a typical field crew member should be able to assess most variables. Greater detail on the second questionnaire and evaluation of the responses is provided in Chap. 3.

The primary conclusions drawn from the responses to the two questionnaires were threefold: (1) the responding countries were in general agreement regarding the most important and feasible variables for assessing biodiversity; (2) harmonization may require that only a few countries introduce substantial numbers of new variables; and (3) a harmonization focus should be agreement on field observation and measurement methods and on field crew expertise (Winter et al. 2008). These conclusions suggested considerable potential for harmonization of biodiversity assessments using NFI data. Therefore, following the assessment of the most important and feasible variables and the assessment of their existing degree of harmonization, WG3 initiated investigation of methods for producing harmonized estimates of biodiversity indicators. This effort focused on constructing bridges that convert estimates based on national definitions to estimates based on reference definitions.

1.5.3 Constructing Bridges

The bridges developed in the third phase can take multiple forms including exclusion of a portion of sample data, complex statistical models to predict missing data, and expert opinion. Crucial factors affecting construction of bridges are the variables and corresponding thresholds in the national definitions under which data are acquired. The nature of these data is the primary factor that distinguishes among three kinds of bridges: *reductive*, *expansive*, and *neutral* bridges. Reductive bridges are appropriate when the national definition is broader in scope than the reference definition. In this case, there is a surplus of national data of which some can be simply excluded. Expansive bridges are necessary when the scope of the reference definition is broader than that of the national definition. In this case, data are missing for estimation based on the reference definition and must be supplied via prediction, imputation, or other method. For construction of expansive bridges, auxiliary variables correlated with the target variable and/or variables are often available to facilitate prediction of missing data. Auxiliary variables may be of many varieties and may be obtained from a variety of sources other than the NFI. When the scopes of the reference and national definitions are the same, neutral bridges are appropriate. Ståhl et al. (submitted) provide a comprehensive discussion of reductive, expansive, and neutral bridging techniques. In summary, bridges are necessary when national definitions deviate from reference definitions. Chapter 5 reports examples of neutral, reductive, and expansive bridges for estimating biodiversity indicators using reference definitions.

1.5.4 Testing Bridges

For the fourth phase, WG3 solicited raw data from NFIs represented among COST Action E43 participants. A database was constructed to serve as a data source for evaluating the utility of the bridges, illustrating the kinds of biodiversity assessments possible with harmonized data, and evaluating the degree to which harmonized assessments are possible in seven categories: forest categories, forest structure, forest age, deadwood, regeneration, ground vegetation and naturalness. Details regarding construction of the database are reported in Chap. 4, and details on construction and testing of bridges are reported in Chap. 5.

1.6 Summary

The investigations of Working Group 3 of COST Action E43 focused on assessing the ability of NFIs to report harmonized estimates of forest biodiversity indicators using NFI data. Four related factors motivated the investigations. Firstly, the importance of forest biodiversity for the economic, environmental, and social well-being of earth's civilizations is gaining wide international acceptance. Secondly, this acceptance has led to numerous international forest sustainability and biodiversity agreements that require periodic reports of estimates of indicators. Thirdly, the ability to report comparable estimates is impeded by the variety of sampling designs, plot configurations, selected variables, and measurement protocols used by the NFIs of different countries. Fourthly, the features of individual NFIs have evolved in response to unique ecological, economic, topographic, and climatic characteristics, and desire of the individual countries to retain the features. The general conclusion of these motivating factors is that apart from substantial standardization of NFIs, the best method for facilitating comparable reporting is to develop harmonization methods.

Working Group 3 undertook a four-phase approach to developing methods for harmonizing estimates of biodiversity indicators using NFI data. The first phase entailed evaluating the importance of biodiversity variables and the feasibility of assessing them using NFI data. The conclusion of this phase was the selection of 17 biodiversity variables that were both important and feasible, grouping of them into seven essential features, and construction of common reference definitions for the variables. The second phase entailed evaluation of the agreement among NFIs with respect to the common definitions and measurement practices. The third phase entailed development of bridges (Ståhl et al. submitted) for converting estimates of forest biodiversity indicators obtained using national definitions to estimates consistent with the reference definitions. The fourth phase entailed construction of a common database of NFI data contributed by NFIs participating in COST Action E43 and testing of reference definitions and bridges developed by Working Group 3.

The following chapters provide details and specific results for the four phases.

References

Achard, F., Eva, H. D., Stibig, H. F., Mayaus, P., Gallego, J., Richards, T., & Malingreau, J. P. (2002). Determination of deforestation rates of the world's humid tropical forests. *Science, 297*, 999–1002.

Adermann, V. (2010). Estonia (National forest inventory report). In E. Tomppo, T. Gschwantner, M. Lawrence, & R. E. McRoberts (Eds.), *National forest inventories: pathways for common reporting (NFI reports section)*. Heidelberg, Springer.

Balmford, A., Bennun, L., ten Brink, B., Cooper, D., Côte, I. M., Crane, P., Dobson, A. P., Dudley, N., Dutton, I., Green, R. E., Gregory, R. D., Harrison, J., Kennedy, E. T., Kremen, C., Leader-Williams, N., Lovejoy, T. E., Mace, G., May, R., Mayaux, P., Morling, P., Phillips, J., Redford, K., Ricketts, T. H., Rodríguez, J. P., Sanjayan, M., Schei, P., van Jaarsveld, A. S., & Walther, B. (2005). The convention on biological diversity's 2010 target. *Science, 307*, 212–213.

Barthlott, W., Kier, G., & Mutke, J. (1999). Biodiversity – the uneven distribution of a treasure. In NNA Reports, Vol. 12, Special issue 2. Forests in focus: Proceedings of biodiversity – treasures in the world's forests. 3–7 July 1998. (ISSN 09 35 - 14 50).

Bosch, P., & Söderbäck, E. (1997). *European environmental state indicators project report*. Stockholm: European Environment Agency, Copenhagen and Swedish Environmental Protection Agency.

Boutin, S., Haughland, D. L., Schieck, J., Herbers, J., & Bayne, E. (2009). A new approach to forest biodiversity monitoring in Canada. *Forest Ecology and Management, 258*, S168–S175.

Brändli, U-B., Bühler., & Zangger, A. (2007). Biodiversität und Waldinventuren. *LFI info 7*, 1–6.

Bundesamt für Naturschutz (BfN). (2002). Daten zur Natur 2002. Münster: Landwirtschaftsverlag. 284 pp.

Bundesnaturschutzgesetz (BnatSchG). (2002). Gesetz über Naturschutz und Landschaftspflege vom 25. März 2002.

Carlowitz HC von. (1713). Sylvicultura oeconomica, oder haußwirthliche Nachricht und Naturmäßige Anweisung zur wilden Baum-Zucht. Leipzig: Braun.

Chadwick, O., Boydak, M., Segura, G., & Bare, B. (1999). Forest organization, management, and policy. In M. L. Hunter Jr. (Ed.), *Maintaining biodiversity in forest ecosystem* (pp. 556–596). Cambridge: Cambridge University Press.

Cienciala, E., Tomppo, E., Snorrason, A., Broadmeadow, M., Colin, A., Dunger, K., Exnerova, Z., Lasserre, B., Petersson, H., Priwitzer, T., Peña, G. S., & Ståhl, G. (2008). Preparing emission reporting systems from forests: use of national forest inventories in European countries. *Silva Fennica, 42*, 73–88.

Convention on biological diversity (CBD). (1992). Convention on biological diversity. Secretariat of the Convention on Biological Diversity, United Nations Environment Programme. http://www.cbd.int/. Accessed Mar 2009.

Convention on biological diversity (CBD). (2007). Article 7, Identification and monitoring. http://www.cbd.int/convention/convention.shtml. Accessed Apr 2008.

Corona, P., & Marchetti, M. (1998). Towards an effective integration of forest inventories and natural resource surveys : the Italian perspective. In M. Hansen & T. Burk (Eds.), Integrated tools for natural resources inventories in the 21st century. Gen Tech Rep. NC-212. pp. 741.

COST. (2009). European cooperation in the field of scientific and technical research. http://www.cost.esf.org/. Accessed May 2008.
COST E43. (2008). Harmonization of national forest inventories in Europe: techniques for common reporting. http://www.metla.fi/eu/cost/e43/index.html. Accessed May 2008.
Council of the European Union. (1992). European Council Directive 92/43/EEC of 21. May 1992 on the conservation of natural habitats and of wild fauna and flora. *Amtsblatt der Europäischen Gemeinschaft Reihe L, 103*(1979), 1–6.
Dinerstein, E., Olson, D. M., Graham, D. J., Webster, A. V., Primm, S. A., Bookbinder, M. P., & Ledec, G. (1995). *Una evaluación del estado de conservación de las ecoregiones terrestres de América Latina y el Caribe. Publicado en colaboración con el Fondo Mundial para la Naturaleza*. Washington, DC: Banco Mundial.
European Economic Community (EEC). (1992). EEC, Council Directive 92/43/EEC of 21 May 1992 on the conservation of natural habitats and of wild fauna and flora, *Official Journal, L 206* (22/07/1992), 7–50.
European Environment Agency (EEA). (2003). *An inventory of biodiversity indicators in Europe, 2002* (p. 42). Copenhagen: European Environment Agency. Technical report number 92/2003.
European Environment Agency (EEA). (2007). *Halting the loss of biodiversity by 2010: Proposal for a first set of indicators to monitor progress in Europe* (p. 182). Copenhagen: European Environmental Agency. Technical report number 11/2007.
European Environment Agency (EEA). (2009). *Progress towards the European 2010 biodversity target* (p. 52). Cepenhagen: European Environmental Agency. Technical report number 4/2009.
Foley, A., DeFries, R., Asner, G. P., Barford, C., Bonan, G., & Carpenter, S. R. (2005). Global consequences of land use. *Science, 309*, 570–574.
Food and Agriculture Organization of the United Nations (FAO). (2005). Global resources assessment 2005. Progress towards sustainable forest management. FAO forestry Paper 147. Rome. 323 p. http://www.fao.org/docrep/008/a0400e/a0400e00.htm. Accessed May 2008.
Franklin, J. F. (1988). Structural and functional diversity in temperate forests. In E. O. Wilson (Ed.), *Biodiversity* (pp. 166–175). Washington, DC: National Academy Press.
Gabler, K., Schadauer, K., Tomppo, E., Vidal, C., Bonhomme, C., & McRoberts, R.E. (submitted). An enquiry on forest areas reported to the Global Forest Resources Assessment 2005 – Harmonization needed? *Forest Science*.
Gallant, A. L., Klaver, R. W., Casper, G. S., & Lanoo, M. J. (2007). Global rates of habitat loss and implications for amphibian conservation. *Copeia, 2007*, 967–979.
Gaston, K.J., & Spicer, J.I. (2004). Biodiversity: An introduction. Blackwell Publishing. 2nd ed.
Global Forest Watch (GFW). (2006). Frequently asked questions. Malden, MA, USA. http://www.globalforestwatch/english/about/faqs.htm. Accessed Jan 2007.
Green, R. E., Balmford, A., Crane, P. R., Mace, G. M., Reynolds, J. D., & Turner, R. K. (2005). A framework for improved monitoring of biodiversity: responses to the world summit of sustainable development. *Conservation Biology, 19*, 56–65.
Gschwantner, T., Schadauer, K., Vidal, C., Lanz, A., Tomppo, E., di Cosmo, L., Robert, N., Duursma, D. E., & Lawrence, M. (2009). Common tree definitions for national forest inventories in Europe. *Silva Fennica, 42*(2), 247–266.
Heath, M. F., Evans, I. E., Hoccom, D. G., Payne, A. J., & Peet, N. B. (Eds.). (2000). *Important bird areas in Europe: Priority sites for conservation* (Birdlife Conservation Series 8, Vol. 2, p. 1600). Cambridge: Birdlife International.
Helsinki Resolution H1. (1993). General guidelines for the sustainable management of forests in Europe. Second ministerial conference on the protection of forests in Europe. 16–17 June 1993, Helsinki/Finland. http://www.mcpfe.org/files/u1/helsinki_resolution_h1.pdf. Accessed Mar 2009.
Holdridge, L. R. (1947). Determination of world plant formation from simple climatic data. *Science, 105*, 367–368.
Holdridge, L. R. (1967). *Life zone ecology*. San Jose: Tropical Science Center.

Hunter, M. L., Jr. (1990). *Wildlife, forests, and forestry: Principles of managing forests for biological diversity* (p. 370). Englewood Cliffs: Prentice-Hall.
Hunter, M. L., Jr. (Ed.). (1999). *Maintaining biodiversity in forest ecosystem* (p. 698). Cambridge: Cambridge University Press.
Iles, K. (1998). Some practical aspects of designing a large inventory. In H. Hansen & T. Burk (Eds.), Integrated tools for natural resources inventories in the 21st century. General Technical Report NC 212. St. Paul: U.S. Department of Agriculture, Forest Service, North Central Research Station. pp 79–85. http://www.ncrs.fs.fed.us/pubs. Accessed Mar 2010.
Intergovernmental Panel on Climate Change (IPCC). (2003). Good practice guidance for land use, land use change and forestry. In J. Penman, D. Kruger, I. Galbally, T. Hiraishi, B. Nyenzi, S. Emmanuel, L. Buendia, R. Hoppaus, T. Martinsen, J. Meijer, K. Miwa, & K. Tanabe (Eds.), IPCC/OECD/IEA/IGES, Hayama, Japan. ISBN 4-88788-003-0. http://www.ipc-nggip.iges.or.jp/public/gpglulucf/gpglulucf_contents.htm. Accessed Apr 2008.
Kleinn, C. (1991). Forest definitions in large-scale inventories. *Allgemeine Forst und Jagdzeitung, 162*(11/12), 201–210.
Köhl, M., & Brassel, P. (1999). La forêt suisse comparée aux autres forêts d'Europe. In P. Brassel & U. B. Brändli (Eds.), *Inventaire forestier national suisse Résultats du deuxième inventaire 1993–1995* (pp. 387–392). Berne, Stuttgart, Vienne, Haupt: Birmensdorf, Institut fédéral de recherches sur la forêt, la neige et le paysage. Berne, Office fédéral de l'environnement, des forêts et du paysage.
Köhl, M., Bachmann, P.,Brassel, P., & Preto, G. (Eds.). (1995). The Monte Verità conference on forest survey designs. "Simplicity versus efficiency" and assessment of non-timber resources. Birmensdorf, Zürich, Swiss Federal Institute for Forest, Snow and Landscape Research (WSL/FNP), Swiss Federal Institute of Technology (ETH), Section of Forest Inventory and Planning. pp. 316.
Köhl, M., Traub, B., & Päivinen, R. (2000). Harmonization and standardization in multi-national environmental statistics – mission impossible? *Environmental Monitoring and Assessment, 63*, 361–380.
Kovac, M., Hocevar, M., & Masvar, R. (2007). A methodological approach for the improvement of biodiversity monitoring and management. In M. Marchetti (Ed.), Monitoring and indicators of biodiversity in Europe – from ideas to operationality. EFI Proceedings No. 51. pp. 526.
Leveque, C. (1994). Environement et diversité du vivant. Pocket Sciences, Collection Explora. pp. 127.
Lindenmayer, D. B., Hobbs, R. J., Montague-Drake, R., Alexandra, J., Bennett, A., Burgman, M., Cale, P., Calhoun, A., Cramer, V., Cullen, P., Driscoll, D., Fahrig, L., Fischer, J., Franklin, J., Haila, Y., Hunter, M., Gibbons, P., Lake, S., Luck, G., MacGregor, C., McIntyre, S., MacNally, R., Manning, A., Miller, J., Mooney, H., Noss, R., Possingham, H., Saunders, D., Schmiegelow, F., Scott, M., Simberloff, D., Sisk, T., Tabor, G., Walke, B., Wiens, J., Woinarski, J., & Zavaleta, E. (2008). A checklist for ecological management of landscapes for conservation. *Ecology Letters, 11*, 78–91.
Lund, H.G. (1986). A primer on integrating resource inventories. Gen. Tech. Dept. W0-49. U.S. Department of Agriculture; Forest Service, Washington, DC: pp. 64.
Lund, H. G. (Ed.). (2006). *Definitions of forest, deforestation, afforestation, and reforestation'*. Gainesville: Forest Information Services. http://home.comcast.net/~gyde/DEFpaper.htm. Accessed Mar 2010.
Mathys, L., Ginzle, C., Zimmermann, N. E., Brassel, P., & Wildi, O. (2006). Sensitivity assessment on continuous landscape variables to classify a discrete forest area. *Forest Ecology and Management, 229*(1/3), 111–119.
Matthews, E., Payne, R., Rohwee, M., & Siobahn, M. (2000). *Pilot analysis of global ecosystems: forest ecosystems* (p. 86). Washington, DC: World Resources Institute.
Meneguzzo, D. M., & Hansen, M. H. (2009). Quantifying forest fragmentation using geographic information systems and forest inventory and analysis plot data. In R. E. McRoberts, G.A. Reams, P.C. Van Deusen, & W.H. McWilliams (Eds.), Proceedings of the eighth annual forest inventory and analysis symposium: 2006 October 16–19; Monterey, California.

General Technical Report WO-79. Washington, DC: U.S. Department of Agriculture, Forest Service:143–148. http://fia.fs.fed.us/symposium/proceedings/. Accessed Mar 2010.
Mikusinski, G., Pressey, R. L., Edenius, L., Kujalah, H., Moilanen, A., Niemela, J., & Ranius, T. (2007). Conservation planning in forest landscapes of Fennoscandia and an approach to the challenge of countdown 2010. *Conservation Biology, 21*(6), 1445–1454.
Ministerial convention on the protection of forests in Europe (MCPFE). (2002). Improved pan-European indicators for sustainable forest management. Liaison Unit Vienna. http://www.mcpfe.org. Accessed Apr 2008.
Ministerial Convention on the Protection of Forests in Europe (MCPFE). (2003). *Improved Pan-European indicators for sustainable forest management as adopted by the MCPFE expert level meeting, 7–8 October 2002* (p. 6). Vienna: MCPFE Liaison Unit.
Montréal Process. (2005). The Montréal Process. http://www.rinya.maff.go.jp/mpci/. Accessed Apr 2008.
Montréal Process. (2006). Criteria and indicators for the conservation and sustainable management of temperate and boreal forests. http://www.mpci.org/rep-pub/1995/santiago_e.html. Accessed Apr 2008.
Nelson, M. D., Lister, A. J., & Hansen, M. H. (2009). Estimating number and size of forest patches from FIA plot data. In R. E. McRoberts, G. A. Reams, P. C. Van Deusen & W. H. McWilliams (Eds.), Proceedings of the eighth annual forest inventory and analysis symposium: 2006 October 16–19; Monterey, California. General Technical Report WO-79. Washington, DC: U.S. Department of Agriculture, Forest Service. pp. 159–164. http://fia.fs.fed.us/symposium/proceedings/. Accessed Mar 2010.
Newton, A. C., & Kapos, V. (2002). Biodiversity indicators in national forest inventories. Kotka IV : Expert consultation on Global F Kotka IV Expert Consultation Global Forest Resources Assessments – Linking National and International Efforts. Background paper 6.4. http://www.fao.org/forestry/foris/webview/forestry/. Accessed Mar 2010.
Nilsson, M., Folving, S., Kennedy, P., Puumalainen, J., Chirici, G., Corona, P., Marchetti, M., Olsson, H., Ricotta, C., Ringvall, A., Ståhl, G., & Tomppo, E. (2003). Combining remote sensing and field data for deriving unbiased estimates of forest parameters over large regions. In P. Corona, M. Köhl, & M. Marchetti (Eds.), *Advances in forest inventory for sustainable forest management and biodiversity monitoring* (Forestry Sciences, Vol. 76, pp. 19–32). Dordrecht: Kluwer Academic Publishers.
Noss, R. F. (1990). Indicators for monitoring biodiversity: a hierarchical approach. *Conservation Biology, 4*(4), 355–364.
O'Neill, R. V., Krummel, J. R., Gardner, R. H., Sugihara, G., Jackson, B., DeAngelist, D. L., Milne, B. T., Turner, M. G., Zygmunt, B., Christensen, S. W., Dale, V. H., & Graham, R. L. (1988). Indices for landscape pattern. *Landscape Ecology, 1*, 153–1626.
Puumalainen, J., Kennedy, P., & Folving, S. (2003). Monitoring forest biodiversity: a European perspective with reference to temperate and boreal forest zone. *Journal of Environmental Management, 67*, 5–14.
Rondeux, J. (1999). Forest inventories and biodiversity. *Unasylva, 196*, 35–41.
Skole, D., & Tucker, C. (1993). Tropical deforestation and habitat fragmentation in the Amazon: satellite data from 1978 to 1988. *Science, 260*, 1905–1910.
Ssymank, A., Hauke, U., Rückriem, C., & Schröder, E. (1998). Das europäische Schutzgebietssystem Natura 2000. *Schriftenreihe für Landschaftspflege und Naturschutz, 53*, 560.
Ståhl, G., Cienciala, E., Chirici, G., Gabler, K., Lanz, A., Vidal, C., Winter, S., McRoberts, R. E., Schadauer, K., & Tomppo, E. (submitted). Bridging national and reference definitions for harmonised provision of forest statistics. *Forest Science*.
Teder, T., Moora, M., Roosaluste, E., Zobel, K., Pärtel, M., Kõljalg, U., & Zobel, M. (2007). Monitoring of biological diversity: a common-ground approach. *Conservation Biology, 21*(2), 313–317.
Tomppo, E., Korhonen, K. T., Heikkinen, J., & Yli-Kojola, H. (2001). Multisource inventory of the forests of the Hebei Forestry Bureau, Heiklong jiang, China. *Silva Fennica, 35*(3), 309–328.

Tomppo, E., Gschwantner, T., Lawrence, M. & McRoberts, R. E. (Eds.). (2010). National forest inventories: pathways for common reporting (pp. 670). Heidelberg, Springer.

Traub, B., Köhl, M., & Päivinen, R. (1997). Simulation study. In: European Commission: study on European forestry information and communication system: report on forest inventory and survey systems. Luxembourg.

United National Environment Programme, Finance Initiative (UNEP FI). (2007). Biodiversity and ecosystem services. October 2007. 12 p. http://www.unepfi.org/fileadmin/documents/CEOBriefing_biodiversity_01.pdf. Accessed Mar 2009.

United Nations Economic Commission for Europe (UNECE). (2003). Declaration by the environment ministers of the region of the United Nations Economic Commission for Europe. Fifth ministerial conference Environment for Europe ECE/CEP/94/Rev.1.

United Nations Economic Commission for Europe (UNECE) and United Nations Food and Agriculture Organization (FAO). (2000). Forest Resources of Europe, CIS, North America, Australia, Japan and New Zealand (industrialized temperate/boreal countries. UN-ECE/FAO contribution to the Global Forest Resources Assessment 2000. Main report. Geneva Timber and Forest Study Papers, 17. United Nations. 445 pp.

United Nations Environment Programme (UNEP). (2001). Indicators and environmental impact assessment: Designing national-level monitoring and indicator programmes, UNEP/CBD/SBSTTA/7/12, Subsidiary body on scientific, technical and technological advice. http://www.biodiv.org/doc/meetings/sbstta/sbstta-07/official/sbstta-07-12-en.pdf. Accessed Mar 2009.

United Nations framework convention on climate change (UNFCCC). (1992). United Nations framework convention on climate change FCCC/INFORMAL-MAL/84 – GE.05-62220 (E) 200705. http://unfccc.int/resource/docs/convkp/conveng.pdf. Accessed Apr 2008.

Vidal, C., Lanz, A., Tomppo, E., Schadauer, K., Gschwantner, T., di Cosmo, L., & Robert, N. (2008). Establishing forest inventory reference definitions for forest and growing stock: a study towards common reporting. *Silva Fennica, 42*, 247–266.

Whittaker, R. H. (1972). Evolution and measurement of species diversity. *Taxon, 21*, 213–251.

Williams, J. (2004). Metrics for assessing the biodiversity values of farming systems and agricultural landscapes. *Pacific Conservation Biology, 10*, 145–163.

Wilson, E. (1992). *The diversity of life* (p. 424). Cambridge: Belknap Press of Harvard University Press.

Wilson, D. B., Cole, F. R., Nichols, J. D., Rudran, R., & Foster, M. S. (Eds.). (1996). *Measuring and monitoring biological diversity: standard methods for mammals* (p. 409). Washington, DC: Smithsonian Institution Press.

Winter, S., Kayser, A., Kolling, T., Pflugmacher, D., Puumalainen, J., von der Wense, H. (2004). Ein Ansatz zur Einbeziehung von Biodiversität und Naturnähe in die Waldinventur und -planung. Forst und Holz:22–26.

Winter, S., Chirici, G., McRoberts, R. E., Hauk, E., & Tomppo, E. (2008). Possibilities for harmonising national forest inventory data for use in forest biodiversity assessments. *Forestry, 81*, 33–44.

Chapter 2
Essential Features of Forest Biodiversity for Assessment Purposes

Susanne Winter, Ronald E. McRoberts, Roberta Bertini, Annemarie Bastrup-Birk, Christine Sanchez, and Gherardo Chirici

Abstract Forest biodiversity assessments may be based on species or taxon groups, structural traits of forest ecosystems and/or biodiversity indicators derived from these variables. Working Group 3 (WG3) of COST Action E43 initially selected 41 candidate biodiversity variables based on current ecological knowledge. The next step entailed construction and distribution of a questionnaire regarding the importance of the candidate variables for assessing forest biodiversity and their feasibility for assessment by national forest inventories (NFI). Responses were received from 22 countries. Analyses of the responses with respect to importance and feasibility resulted in further selection of 17 biodiversity variables that were

S. Winter (✉)
Department für Ökologie und Studienfakultät für Forstwissenschaft und Ressourcenmanagement, Technische Universität München, Germany
e-mail: winter@wzw.tum.de

R.E. McRoberts
Forest Inventory and Analysis, Northern Research Station, USDA Forest Service, USA
e-mail: rmcroberts@fs.fed.us

R. Bertini
Università degli Studi di Firenze, Italy
e-mail: bertini.roberta@gmail.com

A. Bastrup-Birk
University of Copenhagen, Faculty of Life Sciences, Forestry and Wood Products, Denmark
e-mail: ab@life.ku.dk

C. Sanchez
Gembloux Agro-Bio Tech, Université de Liège, Belgium
e-mail: sanchez.c@fsagx.ac.be

G. Chirici
Università degli Studi del Molise, Italy
e-mail: gherardo.chirici@unimol.it

G. Chirici et al. (eds.), *National Forest Inventories: Contributions to Forest Biodiversity Assessments*, Managing Forest Ecosystems 20,
DOI 10.1007/978-94-007-0482-4_2, © Springer Science+Business Media B.V. 2011

then grouped into seven essential biodiversity features: forest categories, forest age, forest structure, deadwood, regeneration, ground vegetation and naturalness. These seven essential features constitute the second level of WG3's 4-level reference framework: (1) concept, (2) essential feature, (3) indicator, and (4) NFI variable. This chapter addresses in detail the analyses of the questionnaire responses, selection of the 17 biodiversity variables, and derivation of the seven essential forest biodiversity features.

2.1 Forest Biodiversity Reference Framework

The investigations of Working Group 3 (WG3) of COST Action E43 were guided by a 4-level reference framework: concept, essential feature, indicator, and NFI variable (Table 1.2). From among the large set of forest management and ecological variables that could be used to assess forest biodiversity, those that can be reasonably assessed by national forest inventories (NFI) must be identified and grouped into a smaller number of categories that are deemed essential for the assessments. To this end, WG3 undertook a systematic approach that included selection of relevant biodiversity variables, evaluation of them with respect to their importance and feasibility for assessment by NFIs, and aggregation of them into essential features. Once the essential features were selected, relevant indicators that can be estimated using NFI variables could then be identified and evaluated with respect to their potential for harmonization. This chapter focuses on the process by which the essential forest biodiversity features were selected.

2.2 Forest Biodiversity Variables

2.2.1 Selecting Forest Biodiversity Variables

The first step in the procedure to select the essential forest biodiversity features was to identify a set of relevant candidate forest management and ecological variables. The selection of these candidate variables was based on information from multiple sources including the Convention on Biological Diversity (CBD 1992; UNEP 2003), the indicators for sustainable forest management established by the Ministerial Conference on the Protection of Forests in Europe (MCPFE 1997, 2003a, b), the Biodiversity Evaluation Tools for European Forests developed in the BEAR project (Larsson et al. 2001), the European Environmental Agency (EEA) Core Set of Indicators for Biodiversity and Nature Protection (EEA 2003), the published forest ecology literature and the expert knowledge of the WG3 participants. On the basis of information from the above cited sources, 41 candidate variables were selected (Table 2.1).

Table 2.1 Candidate variables for assessing forest biodiversity

Variable	Description
Bird species	Number and list of bird species or taxon groups
Bryophyte species	Number and list of bryophyte species or taxon groups
Fungal species	Number and list of fungal species or taxon groups
Herb and grass species	Number and list of herb and grass species or taxon groups
Invertebrate species	Number and list of invertebrate species or taxon groups
Lichen species	Number and list of epiphytic lichen species or taxon groups
Other woody species	Number and list of other woody species or taxon groups
Shrub species	Number and list of shrub species or taxon groups
Tree species	Number and list of tree species or taxon groups
Vertebrate species	Number and list of vertebrates species or taxon groups
Big logs	Lying deadwood with a minimum diameter of 10 cm (the threshold definition is based on the experience acquired in several NFI)
Dead parts on living trees	Potential microhabitats at living trees such as dead branches or crown parts
Decay class	Decay level of the deadwood on the basis of standard definitions of decomposition processes
Deadwood length	Length of lying deadwood
Small logs	Lying deadwood (based on the experience acquired in several NFI the threshold is: minimum diameter smaller than 10 cm)
Snags	Standing deadwood (entire or broken part of dead trees)
Deadwood species	Number of deadwood species or species groups
Stumps	Part of the stem close to the tree roots
Forest category	Classification of forest on the basis of ecological based standardised system of nomenclature (such as EUNIS or BEAR systems)
Naturalness	Similarity of the current forest composition and structure with the natural situation
Information on forest management system	Information regarding silvicultural system (i.e. clearcut system, selection system, shelterwood system, coppice system)
Information on disturbances/damages	Information regarding level of recent man-induced disturbances
Occurrence of microsites	Information regarding presence, quantity and type of microsites as potential microhabitats (such as anthills, rocks accumulation, small humid areas and individual trees features like nesting wholes, crown breakage (Winter and Möller 2008)

(continued)

Table 2.1 (continued)

Variable	Description
Ecotones of microsites	Evaluation of presence, quantity or quality of ecotones (i.e.: by plot partitioning or line intersect sampling)
Regeneration area	Forest area regenerating with forest tree species
Regeneration species	Evaluation of tree species regenerating
Regeneration type	Evaluation of origin of regeneration (natural, planted, seeded)
Shrub height	Evaluation of shrub height
Soil moisture	Evaluation of soil moisture according to national or international standards of classification
Organic layer type	Evaluation of organic component of soil according to national or international standards of classification; mineral versus organic layers
Soil type	Evaluation of soil type according to national or international standards of classification
Development phase	Development phases or stages classifying the natural life cycle
Horizontal structure	Evaluation of the horizontal structure of trees and relative spatial pattern (single trees, groups of trees, etc.)
Vertical structure	Evaluation of the forest layer structure (one, two, more than two layers)
Tree age	Evaluation of age of trees
Tree crown length	Evaluation of crown length
Tree diameter	Evaluation of tree diameter at breast height
Tree health status	Evaluation of vitality or health status on the basis of crown (discoloration, transparency, etc.) or other parts of trees
Tree height	Evaluation of the tree height
Tree infections	Evaluation of the number of trees infected by fungi or other biotic damages including damages by game
Veteran trees	Evaluation of the presence of very old trees

2.2.2 The Importance and Feasibility of Forest Biodiversity Variables

The second step in the procedure consisted of constructing a questionnaire regarding the importance of the candidate variables for assessing forest biodiversity and their feasibility for assessment by NFIs. The questionnaire was made available online to the NFIs of all countries participating in COST Action E43. For each of the 41 candidate variables, the questionnaire included eight questions with predefined multiple choice responses and two questions with unspecified answers (Table 2.2). The experts who responded to the questionnaire had considerable NFI and biodiversity experience and were officially authorized by their countries to complete the questionnaire.

Table 2.2 Questions included in the biodiversity questionnaire for each of the 41 candidate variables

	Question	Possible responses	Description
1	Is the biodiversity feature important as indicator of forest biodiversity?	High importance Moderate importance Low importance	Subjective evaluation of the contribution of the biodiversity feature for the overall assessment of forest biodiversity. Medium refers to average contribution, high refers to essential feature candidates, low refers to an importance clearly lower than the average
2	How feasible is the monitoring of the biodiversity feature by NFI?	High feasibility Moderate feasibility Low feasibility	Assessed evaluation of the total amount of resources needed to incorporate the biodiversity feature in basic field protocols of traditional inventories
3	Is the biodiversity feature currently assessed in your country NFI?	Yes No	Indication whether the biodiversity feature is used or not in field activities of NFI of each country
4	What is the unit used to assess the biodiversity feature?	Open answer	Information regarding the unit used for measuring the biodiversity feature
5	Is the biodiversity feature assessed for all species/types or just for a part of it?	All Selection	Indication whether the biodiversity feature is assessed for all the investigated population or just for a sub-sample (i.e. in a pre-edited list)
6	Which is the source of information?	Sampling plot forest inventory Compartment forest inventory Research Other sources	Indication whether the biodiversity feature is assessed in the full implementation of (e.g. plot or standwise) NFI in the field phase, or if the biodiversity feature is assessed within research or experimental field tests just in selected areas
7	For which kind of land use the biodiversity feature is assessed?	Forest and other woody land Forest only Other woody land only Part of forest and/or other woody land	Indication whether the biodiversity feature is assessed for all population of forest and other wooded land sampling units or just in a sub-sample of it. Definitions of forest and tree may refer to Vidal et al. (2008)
8	What is the assessment method?	Measured Visual estimation Derivation/calculation Determination	Indication whether the biodiversity feature is assessed in the field work by measuring or visual estimation or mathematical derivation by other biodiversity features or proxy biodiversity features. Determination is for those biodiversity features assessed on the basis of pre-edited lists (typically for species)

(continued)

Table 2.2 (continued)

	Question	Possible responses	Description
9	What is the time series of the biodiversity feature in your NFI?	Open answer	Number of years of available comparable data (i.e.: if a NFI is carried out every 5 years and the biodiversity feature was acquired for 2 inventories, the time series is 10 years long)
10	What level of expertise is needed?	No expert Special training Expert	"No expert" refers to field staff usually devoted to field work in the NFI with ordinary forestry background and assessment skills. "Special training" refers to field staff with special training. "Expert" refers to staff with specialized education (e.g. lichenologists for epiphytic lichens, entomologists, soil scientists)

Responses to the questionnaire were received from 22 countries (21 European: Austria, Belgium, Cyprus, Czech Republic, Denmark, Estonia, Finland, France, Germany, Greece, Hungary, Italy, Lithuania, Norway, Portugal, Slovak Republic, Slovenia, Spain, Sweden, Switzerland, United Kingdom) and the Forest Inventory and Analysis (FIA) programme of the United States of America (USA)). No responses were received from countries such as Iceland, Ireland, Latvia that joined COST Action E43 at dates subsequent to the distribution of the questionnaire.

Generally, the more northern and central European countries already include more of the 41 candidate biodiversity variables in their NFIs than did Atlantic and Mediterranean countries (Fig. 2.1). Exceptions were United Kingdom and Spain which both include a large number of relevant biodiversity variables in their NFIs. All 22 responding countries already monitor at least 40% of the 41 biodiversity variables, and 17 countries already monitor at least 50% of the variables. Sweden has the most complete NFI for biodiversity assessment with 91% of the biodiversity variables already assessed, followed by the Slovak Republic with 85%, and the Czech Republic, Finland and Spain with 76% each. Deadwood in the form of big logs and snags is assessed by all responding countries with the exception of Hungary and Portugal. The Slovak Republic, Spain and Switzerland assess all eight of the questionnaire biodiversity variables related to deadwood. All countries assess tree species diversity; all countries except Germany and Hungary assess shrubs or other wooded species; 11 countries assess herbs and grasses; nine countries assess lichens; and six countries assess bryophytes. All responding countries can provide NFI information on tree age or veteran trees with the exception of Switzerland and the USA. All responding countries acquire some kind of information on forest management. Also, assessment of soils is common among NFIs, whereas assessment of fauna-related biodiversity variables is rare;

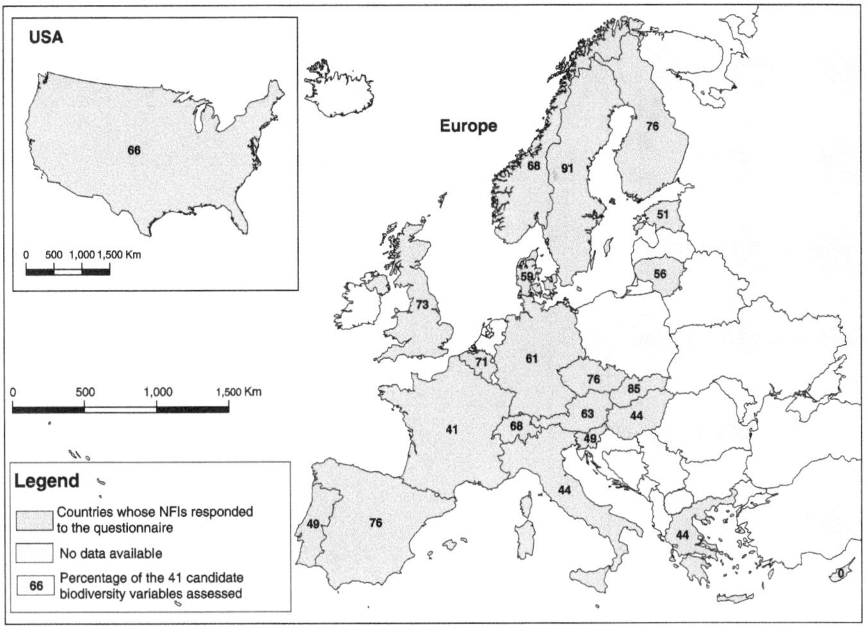

Fig. 2.1 Countries whose NFIs responded to the questionnaire and the percentages of the 41 candidate biodiversity variables that their NFIs assess

only Germany records information on vertebrates and only Lithuania records information on birds.

Responses to the first three questions (Table 2.2) were used to evaluate the candidate variables with respect to their ecological importance and their technical feasibility for monitoring via NFIs. Analyses of responses to the other questions are reported in Chap. 3. Each NFI response for each of the 41 variables was assigned to one of three categories: low, moderate, or high. Variables assigned to the low category were assessed by the country NFI as less than average with respect to importance or feasibility; variables assigned to the moderate category were assessed as mid-level with respect to importance or feasibility; and variables assigned to the high category were assessed as important or feasible variables for assessing biodiversity. The assessment of importance was based on the utility of the variable for describing quantitative or qualitative aspects of forest biodiversity. The assessment of feasibility was based on the total resources in terms of manpower, time, knowledge, and both initial and current costs necessary to incorporate the biodiversity variable into the country's NFI.

Approximately two-thirds of the 41 candidate variables included in the questionnaire were evaluated as very important for monitoring by NFIs, whereas only approximately one-third were evaluated as very feasible. Most other biodiversity variables were evaluated as moderately important and feasible with only a few variables evaluated as less important or less feasible (Table 2.3). Variables evaluated

Table 2.3 Importance and feasibility of candidate biodiversity variables

	Importance				Feasibility			
Variable	Countries responding	High (%)	Medium (%)	Low (%)	Countries responding	High (%)	Medium (%)	Low (%)
Invertebrate species	15	93	7	0	15	7	20	73
Fungal species	14	93	0	7	14	14	43	43
Lichen species	18	83	11	6	18	11	17	72
Veteran trees	17	82	18	0	18	66	28	6
Shrub species	20	80	15	5	20	45	40	15
Tree species	22	77	23	0	22	50	36	14
Big logs	21	76	24	0	21	33	62	5
Snags	20	75	25	0	20	45	50	5
Herb and grass species	18	72	17	11	18	12	44	44
Bird species	17	71	29	0	16	13	25	62
Naturalness	14	71	29	0	14	57	29	14
Development phase	16	69	19	12	17	53	47	0
Forest category	22	68	27	5	22	54	32	14
Other woody species	21	67	24	9	20	30	50	20
Trees age	20	65	35	0	19	21	42	37
Decay class	19	63	37	0	19	47	42	11
Horizontal structure	15	60	33	7	15	60	33	7
Vertical structure	16	56	31	13	17	70	24	6
Microsites	15	53	47	0	15	40	53	7

Information on recent disturbances/damages	19	53	37	10	20	55	40	5
Ecotones of Microsites	14	50	43	7	14	36	36	28
Vertebrate species	16	50	37	13	15	13	27	60
Information on forest management system	22	50	36	14	22	64	36	0
Soil type	20	50	35	15	20	25	35	40
Bryophyte species	15	47	53	0	14	14	43	43
Soil organic layer type	19	47	42	11	19	37	47	16
Deadwood species	22	45	37	18	22	36	59	5
Soil moisture	18	45	22	33	17	30	35	35
Regeneration species	18	44	50	6	18	39	44	17
Regeneration area	19	37	58	5	19	32	57	11
Tree infections	19	37	47	16	20	30	60	10
Tree diameter	22	36	59	5	22	41	50	9
Small logs	20	35	50	15	19	48	26	26
Stumps	18	33	61	6	18	44	44	12
Tree health status	18	33	56	11	19	37	58	5
Dead parts on living trees	18	33	50	17	18	61	33	6
Deadwood length	17	29	59	12	17	29	71	0
Regeneration type	18	28	61	11	18	72	22	6
Shrub height	18	28	33	39	18	50	39	11
Tree height	22	18	68	14	22	27	50	23
Tree crown length	19	16	37	47	19	37	47	16

(% refers to percentage of refers to percentage of responding countries)

as very important are reported below with those also regarded as very feasible reported using italics:

- nine variables related to the number of species (*trees, shrubs*, bryophytes, fungi, herbs and grasses, invertebrates, lichens, and other woody plants);
- three deadwood variables (*snags, decay class* and big logs);
- three forest structure variables (*development phases, horizontal* and *vertical stand structure*);
- two individual tree attribute variables (*veteran trees* and age);
- two variables related to microsites (*occurrence of microsites* and their ecotones);
- two management variables (*information on forest management system* and *information on recent disturbances/damages*);
- *forest category* as it relates to the classification of forests on the basis of an ecological-based standardised system of nomenclature, and knowledge about the organic layer type;
- *forest naturalness*

The third step in the procedure was to combine the responses from individual country NFIs to obtain overall assessments of importance and feasibility for each candidate variable.

2.2.3 Ranking Biodiversity Variables

Based on the aggregation of the questionnaire responses, each of the 41 candidate forest biodiversity variables received an overall evaluation of its importance and feasibility using three measures.

- **Modal value**: Nominal values were assigned to each of the importance and feasibility categories: 1 for low, 2 for mid-level, 3 for high. The modal value is the nominal value associated with the greatest number of responses (Bühl and Zöfel 1999).
- **Index$_1$**:

$$Index_1 = \frac{3 * n_{high} + 2 * n_{moderate} + n_{low}}{n_{responses}} \quad (2.1)$$

  where n_{high}, $n_{moderate}$, and n_{low} were the numbers of high, moderate and low responses, respectively, for each question, and $n_{responses}$ was the total number of responses; values of $Index_1$ ranged between 1 and 3.
- **Index$_2$**:

$$Index_2 = \frac{n_{high} + \dfrac{n_{moderate} + n_{low}}{2}}{n_{responses}} \quad (2.2)$$

  where the definitions were the same as for $Index_1$; values of $Index_2$ ranged between 0 and 1.

Table 2.4 Importance and feasibility thresholds and classes

Class	Modal value	Thresholds for $Index_1$	Thresholds for $Index_2$	Importance of biodiversity feature	Feasibility of biodiversity feature
1	3	>2.33	>0.66	Very important	Very feasible
2	2 and 2.5	1.66–2.33	0.33–0.66	Moderately important	Moderately feasible
3	1 and 1.5	1.00–1.66	0.00–0.33	Less important	Less feasible

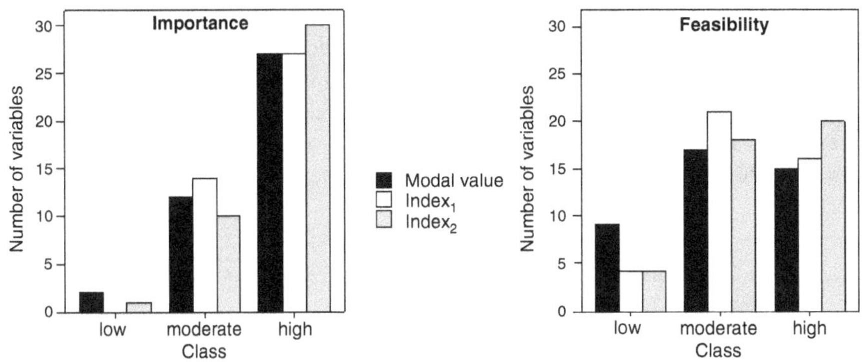

Fig. 2.2 Overall importance and feasibility of candidate variables for monitoring biodiversity in European forests

Each of the 41 candidate biodiversity variables was then assigned to an importance class and to a feasibility class on the basis of selected thresholds (Table 2.4).

The three combined evaluation indices (Modal value, $Index_1$, and $Index_2$) produced nearly the same results (Fig. 2.2).

As a means of evaluating the overall suitability of variables for assessing forest biodiversity by NFIs, the measures of importance and feasibility were combined using three indices:

- **Modal sum**:

$$Modal\,sum = m_1 + m_2 \tag{2.3}$$

  where m_1 = modal value of responses to questions of importance, and m_2 = modal value of responses to questions of feasibility; values of modal sum ranged between 2 and 6.

- **Combination$_1$**:

$$Combination_1 = \frac{Index_{1,importance} + Index_{1,feasibility}}{2} \tag{2.4}$$

  where $Index_{1,importance}$ and $Index_{2,feasibility}$ are as defined in Sect. 2.2.3; values of $Combination_1$ ranged between 1 and 3.

• **Combination$_2$**:

$$Combination_2 = \frac{Index_{2,importance} + Index_{2,feasibility}}{2} \tag{2.5}$$

where $Index_{2,importance}$ and $Index_{2,feasibility}$ were as defined in Sect. 2.2.3; values of $Combination_2$ ranged between 0 and 1.

All 41 forest biodiversity candidate variables were then assigned to suitability classes based on the values of the combined indices (Table 2.5).

The combined data analyses showed that most of the 41 variables were evaluated as at least moderately suitable for assessing forest biodiversity (Fig. 2.3).

Only two variables, bryophyte species and crown length, were deemed less suitable for biodiversity assessments using NFI data. The bryophyte species were assessed as moderately important for reporting biodiversity and moderately to less feasible for field assessment (Table 2.3). The crown length variable was

Table 2.5 Combined index classification of the importance and feasibility for candidate biodiversity variables

Modal sum	Thresholds for Combination$_1$	Thresholds for Combination$_2$	Suitability for forest biodiversity monitoring by NFI
5, 5.5 and 6	>2.33	>0.66	Very suitable
3.5, 4 and 4.5	>1.66–2.33	>0.33–0.66	Moderately suitable
2, 2.5 and 3	1–1.66	0–0.33	Less suitable

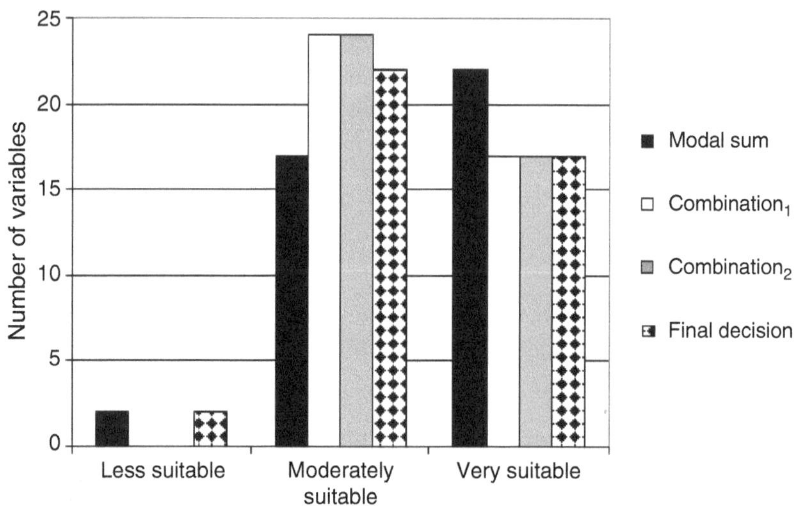

Fig. 2.3 Distribution of the biodiversity variables by the three suitability categories

evaluated as least suitable for biodiversity assessments using NFI data. Most other plant and animal groups were evaluated as moderately suitable for NFI on the basis of high importance but low feasibility. For example, invertebrate species were considered as a very important biodiversity variable by 93% of respondents, but 73% assessed its feasibility as low. A second example is lichen species for which 83% of the experts responded that it was important but 72% evaluated its feasibility as low.

2.3 The Essential Forest Biodiversity Features

The results of combining indices of importance and feasibility were that 17 biodiversity variables were classified as very suitable. To simplify the harmonization analyses (Chaps. 3 and 5), 13 of the 17 variables were aggregated into seven groups which were then designated as essential features of forest biodiversity (Table 2.6). Of the four remaining variables, information on forest management system and recent disturbances were not selected because of their widely varying assessment methods among countries, suggesting a low potential for harmonization. Microsites and dead parts of living trees were not selected because they were generally not assessed by NFIs. However, lack of assessment for the latter two variables provides an opportunity for construction and widespread adoption of a common reference definition before individual NFIs construct their own differing national definitions, thus eliminating the need for harmonization.

Table 2.6 The Working Group 3 essential features of forest biodiversity

Biodiversity variable	Number of countries that assessed the variable	Essential feature of forest biodiversity
Forest category	19	Forest categories
Development phase	11	Forest structure
Horizontal structure	10	
Vertical structure	16	
Trees species	21	
Tree diameter	21	
Big logs	19	Deadwood
Snags	17	
Decay class	15	
Regeneration type	19	Regeneration
Veteran trees	12	Forest age
Shrub species	16	Ground vegetation
Naturalness	10	Naturalness

2.4 Discussion

NFI participants seldom responded to all questions in the questionnaire. Thus, suitability assessments for bryophyte and fungi species, microsites and their ecotones, and forest naturalness are considered less reliable because they received few responses. However, the suitability assessments for tree diameter, height and species, forest category, deadwood species, and information on the forest management system are considered highly reliable for European countries and the USA because they received responses for both importance and feasibility from all 22 countries. Participants may reasonably be assumed to have provided responses mainly for biodiversity variables used by their own NFIs and for which they have assessment experience.

Additionally, participants may have responded more frequently to questions about biodiversity variables they regarded as having high or moderate relevance for biodiversity. Of the 41 candidate variables, the ranking analysis based on the modal value classified 26 of them as highly important for biodiversity monitoring by NFIs; only two variables were evaluated as less important. A possible confounding issue is that only a few biodiversity variables were classified as low with respect to importance or feasibility. This phenomenon may possibly be attributed to three factors. First, the reasons for selecting the biodiversity variables for inclusion in the questionnaire are ecologically based and are well-documented in the literature. Second, the participants may not have responded when they judged the importance or feasibility of a biodiversity variable as low or when they were not sure about its importance (13 variables were only assessed by 13–17 of the participating 22 countries). Third, some biodiversity variables such as microsites may be unfamiliar to participants whose countries do not assess them. However, increasing knowledge of forest ecosystems and their biodiversity and natural structure could change this judgment in the future.

Beyond these possible limitations, information obtained from the questionnaire and the subsequent analyses clearly showed that currently most of the 41 candidate forest biodiversity variables are monitored by the 22 NFIs that responded to the questionnaire. With respect to the number and type of questionnaire biodiversity variables, the FIA programme of the U.S. Forest Service collects an average amount of information. In addition, most of the countries already collect information on nearly all the essential biodiversity features (Table 2.6).

The importance and feasibility analyses clearly confirmed that NFIs prefer biodiversity variables based more on forest structure indicators such as vertical, horizontal and tree compositional diversity or deadwood than on direct biological diversity measures of animals such as birds and invertebrates or vegetal life forms such as bryophytes, fungi, herbs, grasses and lichens. In general, biota biodiversity variables were evaluated as important but not feasible because their assessment is excessively intensive relative to time, cost, and necessary expertise. However, the thematic resolution of information on structural indicators is much coarser than the fine resolution information associated with individual species and their ecological niches. Thus, biodiversity assessments based on structural variables cannot produce estimates that are fully comparable to results obtained from direct measures of biodiversity.

The responses to the other eight questions indicated in Table 2.2 and to a second questionnaire on methods used by NFIs to assess variables associated with the essential features of forest biodiversity are reported in Chap. 3.

References

Bühl, A., & Zöfel, P. (1999). *SPSS Version 8 – Einführung in die moderne Datenanalyse unter windows*. Kevelaer: Addison-Wesley Scientific Computing.

CBD. (1992). Convention on biological diversity. Secretariat of the Convention on Biological Diversity, United Nations Environment Programme.

EEA. (2003). Core set of indicators. http://ims.eionet.eu.int/Topics/BDIV. Accessed Apr 2006.

Larsson, T.-B., Svensson, L., Angelstam, P., Balent, G., Barbati, A., Bijlsma, R.-J., Boncina, A., Bradshaw, R., Bücking, W., Ciancio, O., Corona, P., Diaci, J., Dias, S., Ellenberg, H., Fernandes, F. M., Fernàndez-Gonzalez, F., Ferris, R., Frank, G., Møller, P. F., Giller, P. S., Gustafsson, L., Halbritter, K., Hall, S., Hansson, L., Innes, J., Jactel, H., Keannel Doppertin, M., Klein, M., Marchetti, M., Mohren, F., Niemelä, P., O'Halloran, J., Rametsteiner, E., Rego, F., Scheidegger, C., Scotti, R., Sjöberg, K., Spanos, I., Spanos, K., Standovár, T., Tømmerås, Å., Trakolis, D., Uuttera, J., Walsh, P. M., Vandekerkhove, K., Watt, A. D., & VenDenMeersschaut, D. (2001). Biodiversity evaluation tools for European forests. A report from the FAIR project "Indicators for monitoring and evaluation of forest biodiversity in Europe" CT97-3575 within the EU Commission RTD Programme. Ecological Bulletin 50.

MCPFE. (1997). Work-programme on the conservation and enhancement of biological and landscape biodiversity in forest ecosystems. Adopted at expert-level by the Third Meeting of the Executive Bureau of the Pan-European Biological and Landscape Diversity Strategy, Geneva, Switzerland, 20–21 Nov 1997.

MCPFE. (2003a). Vienna Declaration and Vienna Resolutions adopted at the Fourth Ministerial Conference on the Protection of Forests in Europe, Vienna Austria, 28–30 Apr 2003.

MCPFE. (2003b). Improved Pan-European indicators for sustainable forest management as adopted by the MCPFE Expert Level Meeting, Vienna, Austria, 7–8 October 2002.

UNEP. (2003). Proposed biodiversity indicators relevant to the (2010) target. UNEP/CBD/SBSTTA/9/INF/26.

Vidal, C., Lanz, A., Tomppo, E., Schadauer, K., Gschwantner, T., Di Cosmo, L., & Robert, N. (2008). Establishing forest inventory reference definitions for forest and growing stock: a study towards common reporting. *Silva Fennica, 42*(2), 247–266.

Winter, S., & Möller, G. (2008). Microhabitats in lowland beech forests as monitoring tool for nature conservation. *Forest Ecology and Management, 255*, 1251–1261.

Chapter 3
Prospects for Harmonized Biodiversity Assessments Using National Forest Inventory Data

Ronald E. McRoberts, Gherardo Chirici, Susanne Winter, Anna Barbati, Piermaria Corona, Marco Marchetti, Elmar Hauk, Urs-Beat Brändli, Jana Beranova, Jacques Rondeux, Christine Sanchez, Roberta Bertini, Nadia Barsoum, Iciar Alberdi Asensio, Sonia Condés, Santiago Saura, Stefan Neagu, Catherine Cluzeau, and Nabila Hamza

Abstract Following selection of the 13 biodiversity variables that were evaluated as both important and feasible for assessment by NFIs and grouping them into essential features, additional information was solicited regarding the degree to which the 13 variables are currently assessed by NFIs. The objective was to evaluate the prospects for harmonized estimates of biodiversity indicators based on these variables. The prospects varied considerably depending on the particular variable and essential feature. The evaluations produced positive harmonization

R.E. McRoberts (✉)
Forest Inventory and Analysis, Northern Research Station,
USDA Forest Service, USA
e-mail: rmcroberts@fs.fed.us

G. Chirici and M. Marchetti
Università degli Studi del Molise, Italy
e-mail: gherardo.chirici@unimol.it; marchettimarco@unimol.it

S. Winter
Department für Ökologie und Studienfakultät für Forstwissenschaft und Ressourcenmanagement, Technische Universität München, Germany
e-mail: winter@wzw.tum.de

A. Barbati and P. Corona
Università degli Studi della Tuscia, Italy
e-mail: barbati.sisfor@unitus.it; piermaria.corona@unitus.it

E. Hauk
BFW, Federal Research and Training Centre for Forests,
Natural Hazards and Landscape, Austria
e-mail: elmar.hauk@bfw.gv.at

U.-B. Brändli
Swiss Federal Institute for Forest, Snow and Landscape Research (WSW),
Schweizerisches Landesforstinventar, Switzerland
e-mail: urs-beat.braendli@wsl.ch

G. Chirici et al. (eds.), *National Forest Inventories: Contributions to Forest Biodiversity Assessments*, Managing Forest Ecosystems 20,
DOI 10.1007/978-94-007-0482-4_3, © Springer Science+Business Media B.V. 2011

possibilities for forest categories and the tree height and diameter variables associated with forest structure. For forest age, possibilities were constrained by lack of common reference definitions. However, possibilities for construction of a common reference definition and bridges to compensate for the differences in estimates resulting from using national and reference definitions were deemed positive. Prospects for regeneration, ground vegetation, and naturalness were less positive because of variability in definitions, assessment methods, measurement thresholds and other factors. Thus, efforts at harmonization for these essential features were constrained to a few variables or a few countries with similar NFI features.

3.1 Introduction

Selection of the 13 biodiversity variables and grouping of them into seven essential features (Table 2.5) was based primarily on the responses to the first three questions of the biodiversity questionnaire (Table 2.2) and are documented in Chap. 2.

J. Beranova
IFER, Institute of Forest Ecosystem Research Ltd, Czech Republic
e-mail: iana.beranova@ifer.cz

J. Rondeux
Gembloux Agro-Bio Tech, Université de Liège, Belgium
e-mail: jrondeux@ulg.ac.be

C. Sanchez
Gembloux Agro-Bio Tech, Université de Liège, Belgium
e-mail: sanchez.c@fsagx.ac.be

R. Bertini
Università degli Studi di Firenze, Italy
e-mail: bertini.roberta@gmail.com

N. Barsoum
Forest Research, Forestry Commission, United Kingdom
e-mail: nadia.barsoum@forestry.gsi.gov.uk

I.A. Asensio
Departamento Sistemas y Recursos Forestales, Centro de Investigación Forestal,
Instituto Nacional de Investigación y Tecnología Agraria y Alimentaria, Spain
e-mail: alberdi.iciar@inia.es

S. Condés and S. Saura
Escuela Técnica Superior de Ingenieros de Montes, Universidad Politécnica de Madrid, Spain
e-mail: ifnbiodiv@mma.es; sonia.condes@upm.es; santiago.saura@upm.es

S. Neagu
ICAS, Forest Research and Management Institute, Romania
e-mail: biometrie@icas.ro

C. Cluzeau and N. Hamza
IFN, French National Forest Inventory, France
e-mail: catherine.cluzeau@ifn.fr; nabila.hamza@ifn.fr

The NFIs of 26 countries, including the FIA programme of the United States of America (USA), responded to the first (Table 2.2) and to the second questionnaire. Responses to Questions 3 (Table 2.2) indicated that each variable is assessed by the NFIs of at least eight countries, but only two variables, tree diameter and tree species, are assessed by the NFIs of all 26 responding countries. Thirteen variables are assessed by the NFIs of 10–21 countries. Most variables are assessed for all relevant plot components; variables assessed only for fewer components include tree age but only for dominant trees and microsites.

Responses to Question 10 indicated considerable lack of agreement among responding NFIs regarding the expertise necessary to assess variables. Eight NFIs responded that a high level of expertise is necessary for most variables, although most NFIs responded that mid-level expertise is necessary but with a tendency toward a higher level. Only Germany and the USA responded that a typical inventory field crew member should be able to assess most variables. As expected, responses to Question 7 indicated that forest land is the category for which most variables are assessed.

Based on responses to Question 8, the NFIs of 14 of the 22 responding countries use all four field assessment methods. Few NFIs use the identify method, and the derive method is limited to variables such as forest type, forest naturalness, and development phases. Only tree age is a derived variable in most NFIs with assessment based on elapsed time since regeneration or counts of tree rings. Visual estimation is the most common method for variables used to assess disturbance/damage, vertical structure, and deadwood decay classes.

The question on thresholds of the second questionnaire was intended to assess possibilities for harmonization. Two examples illustrate the diversity of thresholds used by NFIs. First, although dbh is assessed by every responding NFI, minimum dbh thresholds range from 0 to 120 mm. Only five countries use the same threshold, 70 mm. Second, the NFIs of 20 countries assess vertical structure using number of tree layers. Most countries use three categories of layers: one, two, and more than two layers. However, the boundaries for the three layers vary considerably; for example, as a lower threshold for the lowest layer, one country uses 0.1 m while another uses 4 m. In addition, one country uses a different boundary for plantations than for other kinds of forest.

Three general conclusions were drawn from the responses to the questionnaire. First, the general consensus on the most ecologically important and technically feasible biodiversity variables bodes well for the possibilities for harmonization. Second, because most NFIs already assess a large proportion of the 13 variables, Europe-wide harmonization would require introduction of substantial numbers of new variables by only a few NFIs. Third, the lack of agreement on assessment methods and necessary crew expertise suggests that harmonization may need to emphasize field operations.

More details on results for individual biodiversity variables are provided in the following sections, one for each of the seven selected essential features (Table 2.5).

3.2 Forest Categories

3.2.1 Background

The term *forest categories* refers to a system of nomenclature for classifying forest and other wooded lands. Usually such categories are adopted for reporting and/or for stratification in different stages of the inventory. The national forest inventories (NFI) of 17 of the 23 countries responding to the forest structure portion of the second questionnaire use a standard system of nomenclature for forest categories (Fig. 3.1).

Because forest categories are usually defined as groupings of similar forest types, a brief discussion of forest types is necessary before addressing harmonization of NFIs with respect to forest category classification schemes. *Forest type* is a very broad term, with several meanings. The Montréal Process (1998) defines forest type as "a category of forest defined by its composition, and/or site factors (locality), as categorized by each country in a system suitable to its situation". Forest type classifications provide a flexible approach for collecting and organising forest information for multiple purposes such as large area analyses of forest productivity, silvicultural treatment planning, sustainable forest management and assessment of the structure and composition of potential natural vegetation.

Forest type classifications differ either in the nature of the forest communities that are the object of the classifications, as exemplified by potential versus actual

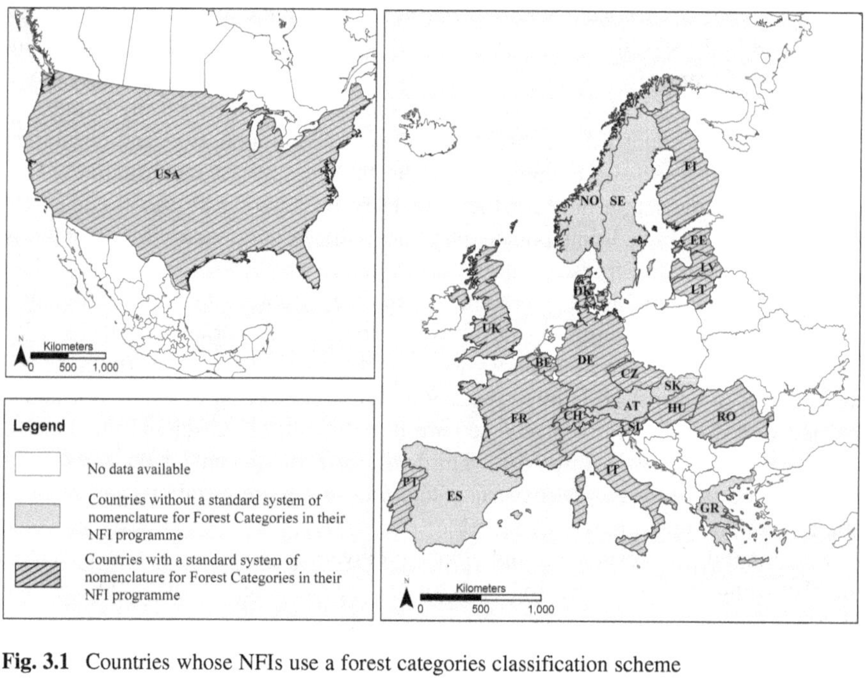

Fig. 3.1 Countries whose NFIs use a forest categories classification scheme

forest vegetation communities, or in the number of classes used, as exemplified by the 377 forest habitat types used by the European Nature Information System (EUNIS) (Davies et al. 2004) or the 699 potential forest vegetation communities used for the European map of natural vegetation (Bohn et al. 2000). In the framework of NFIs, a forest type classification may be used as a component of a sampling strategy aimed at improving the assessment of forest condition over a large area or it may be used to report amounts of selected forest resources for commercial planning purposes. Forest types partition a forest area into a discrete number of smaller and more ecologically homogeneous components and facilitate analysis, interpretation, and reporting of forest data.

The use of a forest type classification is often recommended when addressing forest biodiversity assessment issues. This approach was suggested in the framework of the pan-European concerted action "Indicators for monitoring and evaluation of forest biodiversity in Europe" (BEAR), which outlined an optimised strategy for the assessment of forest biodiversity based on natural and anthropogenic determinants and related indicators (Larsson et al. 2001). The proposed forest type classification was based on 33 forest types representing a heterogeneous mixture of actual and potential forest vegetation with marked differences in the relative importance of individual types. The need for a better assessment of European forest conditions has prompted a long process of revision of the BEAR scheme. One result is a new proposal for *European Forest Types* (*EFT*) for classification of European forests based on a two-level system of nomenclature: 76 forest types grouped into 14 primary forest categories (EEA 2006; Barbati et al. 2007). This classification was proposed as a reference scheme for reporting harmonized NFI estimates of forest biodiversity indicators. The forest type level of the EFT system can be considered the tool to create a bridge between local forest category systems of nomenclature and the 14 reference forest categories proposed for the EFT system.

3.2.2 Forest Categories Assessment Within NFIs

The NFIs of most responding countries use a standard system of nomenclature for forest categories (Fig. 3.1). However, in countries whose NFIs do not include a forest category scheme, forest classification is still commonly used for stratification, reporting or both. The NFIs of all responding countries provided additional information on the characteristics of their forest category classification including whether the objective was actual or potential forest vegetation, the primary classification variables, and purpose of the classification. Potential forest vegetation reflects the ecological potential of forest sites and the biological diversity of potential forest communities across the country. Potential forest communities are not necessarily aligned with actual forest vegetation as surveyed by NFIs, but rather, they provide a consistent and reliable basis for assessment for biodiversity features such as similarity to natural forest conditions. The forest category schemes of 11 NFIs classify actual vegetation with no direct relationship to potential

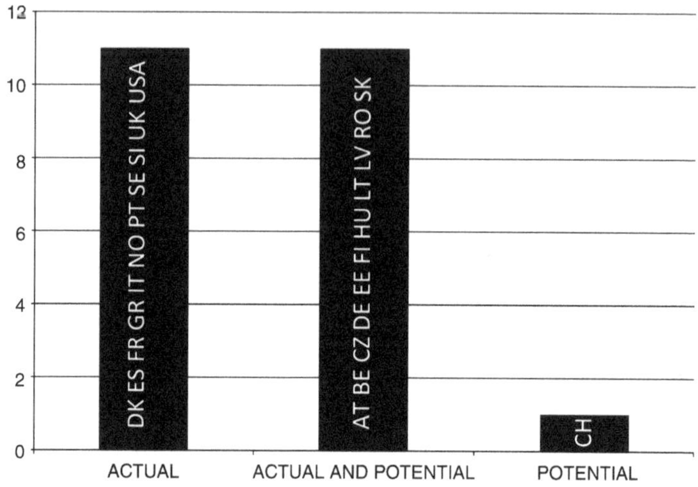

Fig. 3.2 Use of actual or potential vegetation in European forest category classification schemes (see the section Abbreviations for country codes)

vegetation, whereas 11 NFIs use schemes covering both actual and potential vegetation (Fig. 3.2). Switzerland is the only country whose forest category classification scheme is based exclusively on potential forest vegetation.

Tree species composition is the primary classification variable for assessing forest categories in all countries, and for most countries it is the only variable. Site variables such as temperature regimes and water availability are used in six countries; altitude, climate and biogeographical areas are considered in five countries; soil information is used by just three countries; and phytosociological vegetation variables are used by three countries. Other classification variables include forest cover, stand age, stand structure, regeneration, and ground vegetation.

Forest category classifications are used by the NFIs of responding countries for four primary purposes: (1) stratification within the sampling design (seven countries); (2) improved NFI reporting such as volume estimates by forest category (eight countries); (3) mapping of tree species and forest community distributions (four countries); and (4) evaluation of forest naturalness by comparing actual and potential vegetation (three countries). The remaining countries do not have a defined purpose for forest categories and may use the information for one or more of the previous purposes (ten countries).

The primary method for classifying forest into forest categories in European NFIs is field work with 10 of 17 countries using this method exclusively. Only one country relies on the interpretation of aerial photography as the sole source of information. Six countries delineate forest categories using aerial photographs; one country uses satellite imagery; and two countries use information available in a variety of GIS layers. The minimum forest area necessary for assessing forest category in the field ranges between 0.05 and 1 ha. When forest categories are mapped, the size of the minimum mapping unit ranges between 0.1 and 400 ha.

3.2.3 *Similarities and Differences in the Use of Forest Category Classifications*

In summary, approximately 75% of the NFIs of responding countries have adopted forest category classification schemes. Classification of forest stands by forest categories is, therefore, a familiar activity for NFI survey teams. The primary similarity among forest category classification schemes is that tree species composition is the main classification variable. The primary difference pertains to whether actual or potential vegetation is used as the reference for the classification.

3.2.4 *Harmonization Possibilities*

The first step in harmonization of European NFIs with respect to forest categories is to construct an international forest categories reference scheme. Such a scheme should be sufficiently comprehensive to accommodate the variability in actual forest conditions at the national level and should rely on classification variables such as tree species composition and site variables that are commonly assessed by NFIs.

For the harmonization test (Sect. 5.2) using data from the common NFI database (Chap. 4), the EFT system of nomenclature at category level was adopted as the international reference scheme. NFI plots were classified according to the EFT system at forest type level on the basis of tree species composition and then classified into forest categories on the basis of their forest types.

3.3 Forest Structure

3.3.1 *Background*

The approach selected by Working Group 3 (WG3) for assessing forest structure relates to the physical organization of forest elements with respect to composition and complexity. Forest structure is one of the most important features of forest biodiversity because it encompasses three-dimensional forest space using a combination of variables representing horizontal, vertical and species composition. For old-growth forests whose structural heterogeneity includes entire life cycles, the effects of natural disturbances, small gaps, dead wood and natural regeneration produce high rankings on the naturalness and biodiversity scales (Bartha et al. 2006; Michel and Winter 2009). In addition, positive correlations have been found between forest structural richness and forest flora and fauna (MacMahon et al. 1981; Winter et al. 2005), particularly bird species and abundance (De Graaf et al. 1998; Schumacher 2006; McRoberts 2009) and saproxylic coleoptera (Winter and Möller 2008). The overall level of natural structural heterogeneity of undisturbed forests depends on the growth conditions of the forest community which, in turn, depend mainly on climate and soil. In general, more northern forest communities, such as in boreal forests, have less complex natural structures than communities in southern forests, such as in mountainous mixed forests in the temperate zone. However, regardless of climatic constraints, more heterogeneous forests, even in boreal regions, are rich in habitats and favour the life conditions necessary for forest specialists as opposed to ubiquitous species (Hanski and Hammond 1995; Winter and Möller 2008).

Most European temperate forests have been shaped by silvicultural management and are not threatened by deforestation (Foley et al. 2005). However, these forests suffer loss of ecological quality by simplification to secondary stands and fragmentation (Noss 1990) and are inhabited by increasing numbers of ubiquitous species and decreasing numbers of forest specialists (Duelli and Obrist 2003). The effects of human influence on forest structure have been quantified using the *hemeroby* concept[1] (Blume and Sukopp 1976; Grabherr et al. 1998) which has also been used to assess forest naturalness (Sect. 3.8, McRoberts et al. submitted, Petriccione 2006).

Currently, innovative forest management guidelines often combine objectives related to both economics and forest naturalness (Zerbe and Kempa 2005; Flade et al. 2004). The primary objective of monitoring forest structure is to assess sustainable forest management (Ciancio et al. 1999). To accomplish this objective, NFIs must collect scientifically sound and comparable biodiversity information so that estimates of biodiversity indicators can be harmonized. One problem is that observations for relevant variables obtained on plots of different sizes are less comparable. Unfortunately, NFI plot sizes in Europe and elsewhere exhibit considerable variability. For example, the Swedish and Austrian NFIs assess forest structure on 300 m^2 plots (Axelsson and Fridman 2005; Schieler and Hauk 2001), whereas the Slovenian and Spanish NFIs use variable radius plots whose sizes range up to 2,000 m^2 (Hočevar and Kovač 2004; Montes et al. 2005). The Walloon NFI assesses stand disturbances and silvicultural types on approximate 4,000 m^2 plots (Rondeux et al. 2005), and the National Institute of Agricultural Research (*INIA*) in Spain assesses biodiversity on 5,000 m^2 plots (Montes et al. 2005). The effects of different NFI plot sizes on estimates of structural indicators such as the Shannon index of tree species and large tree dbh are well documented (Sect. 5.8, McRoberts et al. 2009, McRoberts et al. submitted). NFIs are assumed to have selected their features, including plot size and configuration, to accommodate their unique forest conditions, topographies, climates, and commercial interests. For example, trees in oak forests in dry Mediterranean regions are naturally widely spaced, but mountainous beech forests with spruce and fir in regions with high annual precipitation naturally have greater tree densities. Therefore, the Spanish NFI requires a larger plot size than NFIs for humid mountainous regions. Standardization with respect to plot sizes and configurations produces statistically comparable data, although ecological

[1] *hemeroby*: from the Greek *hemeros* meaning cultivated, tamed, or refined (Jalas 1955); a measure that integrates the effects of past and present influence on ecosystems (Sukopp et al. 1990, Kowarik 1999).

relevance and reliability could be reduced. Thus, harmonization, which preserves ecological relevance, may be the better route to comparability.

One of the lessons learned from the literature is that NFI data should be used to estimate indicators that aggregate multiple biodiversity variables (especially of taxon groups) and that are easy to measure. A complementary approach is to concentrate on a small number of species that are representative, important and/or endangered and whose habitat requirements are narrow but yet broad enough that variables related to the species can serve as indicators. For example, the spotted owl *Strix occidentalis caurina* is an indicator species for high structural heterogeneity in old-growth stands in the western USA (Brodie et al. 2007). However, information for the indicator species approach would be difficult to harmonize for large areas.

Forest structure is an essential biodiversity feature that aggregates information for multiple variables: development phases, tree species composition and vertical and horizontal structure. To assess possibilities for harmonization, the second WG3 questionnaire solicited information from NFIs on multiple forest structure variables: (1) tree dbh, (2) tree height, (3) tree species, (4) number of trees/ha, (5) coordinates of the measured trees, (6) distance to the nearest neighbour tree, (7) tree social position, (8) number of crown layers, (9) species distribution by crown layer, (10) area with gaps, (11) information on forest edges and (12) stage of development.

3.3.2 Forest Structure Assessment within NFIs

Responses to the second WG3 questionnaire on NFI assessments of variables related to essential biodiversity features were received from 26 countries. The responding countries currently include many variables related to forest structure in their NFIs (Fig. 3.3), an unsurprising result because monitoring of forest structure for purposes of estimating wood use, economic turnover and profit is among the main objectives of all NFIs, regardless of any focus on monitoring forest biodiversity. Thus, harmonization of estimates of indicators of forest structure would seem to be easy because all countries record information on forest structure. Simultaneously, harmonization could be difficult because each country has a long history of using its own unique variables and indicators, and even slight changes in approaches or variables may adversely affect long time series of inventory data. Tree species, tree diameter, tree height and stand density (trees per unit area) are the only four variables that are assessed by the NFIs of at least 25 of the 26 countries (Fig. 3.4). However, more than half the NFIs acquire information on the other variables with the exception of forest edges (12 countries) and gaps (eight countries). These results suggest that prospects are excellent for harmonizing indicators of forest structure at a continental scale.

Most NFIs in northern and eastern Europe assess 80% of the variables on forest structure, whereas NFIs in southern and western countries assess only approximately half the variables; only Sweden's NFI assesses all the variables. Information for forest structure variables is generally available with more than 90% of responding NFIs assessing more than 50% of the variables.

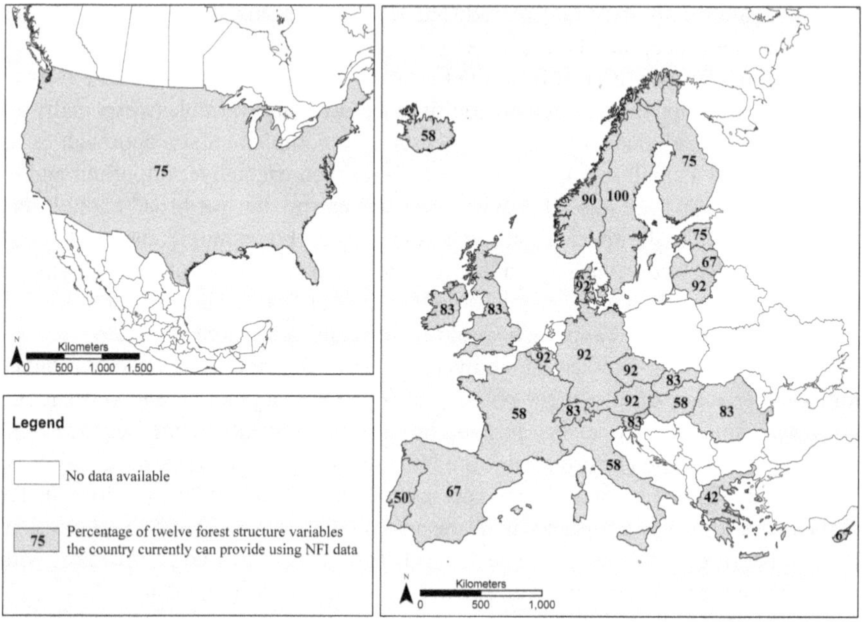

Fig. 3.3 Percentages of 12 forest structure variables assessed by the NFIs responding to the WG3 questionnaire

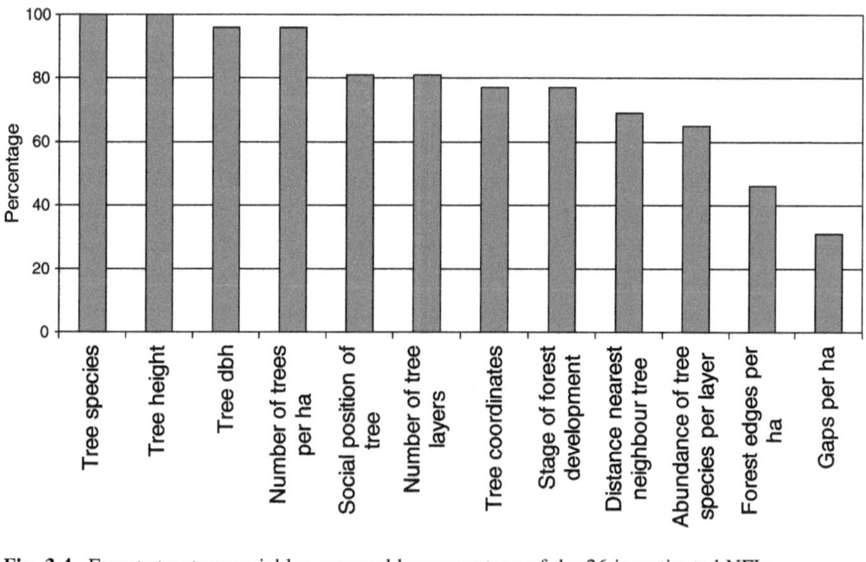

Fig. 3.4 Forest structure variables assessed by percentage of the 26 investigated NFIs

3.3.3 *Similarities and Differences in the Use of Forest Structure Indicators*

To assess the possibilities for harmonization in greater detail, methods based on the most frequently assessed variables (tree species, dbh and height; density; social position and layers) were described and compared among countries. Information for gaps, development phases and edges were not used because of the small number of countries whose NFIs assess them.

3.3.3.1 Tree Species

Tree species is assessed by all participating NFIs, and because the assessments are conducted in the same manner and recorded mostly in the same detail, the possibilities for harmonization are exceptionally good. When species are only recorded by species groups or genera, such as a group including only deciduous trees with short life spans, three problems arise: (1) gamma biodiversity, the absolute number of tree species in a region, cannot be evaluated, (2) estimates of alpha diversity at the plot level lose resolution because the species of individual trees in a species group cannot be distinguished and (3) proportions of native and non-native tree species can only be approximated for the estimation of beta diversity indices. Despite this minor inconsistency, indicators of the diversity of tree composition are among biodiversity indicators that WG3 tested with NFI data (Sect. 5.3).

For countries that record individual tree species, indicators based on native and non-native species were tested (Sect. 5.3). The assessment of nativeness is based on the EEA (2006) forest categories which, in turn, consider the natural range of each species. The NFIs of 13 countries already estimate the percentage of non-native tree species, and an additional country has announced that it will include this estimate in the future.

Changes in tree species composition can be estimated using time series with eight countries having time series data ranging in length from 7 to 84 years. Finland has the longest time series, 84 years, although Cyprus has gathered tree species information for 54 years, and Slovenia has conducted four inventories since 1987. For reporting biodiversity changes, harmonization must address differences in time intervals between measurements.

3.3.3.2 Tree dbh and Height

Although countries have comparable objectives in their use of tree dbh and height for calculating volumes of individual trees, the minimum thresholds for measuring these variables vary considerably. For dbh, the range is 0–120 mm with a median of 64 mm. Five countries use the same minimum dbh threshold of 70 mm which is closest to the median (Fig. 3.5); only one country does not record information on the diameter of the trees.

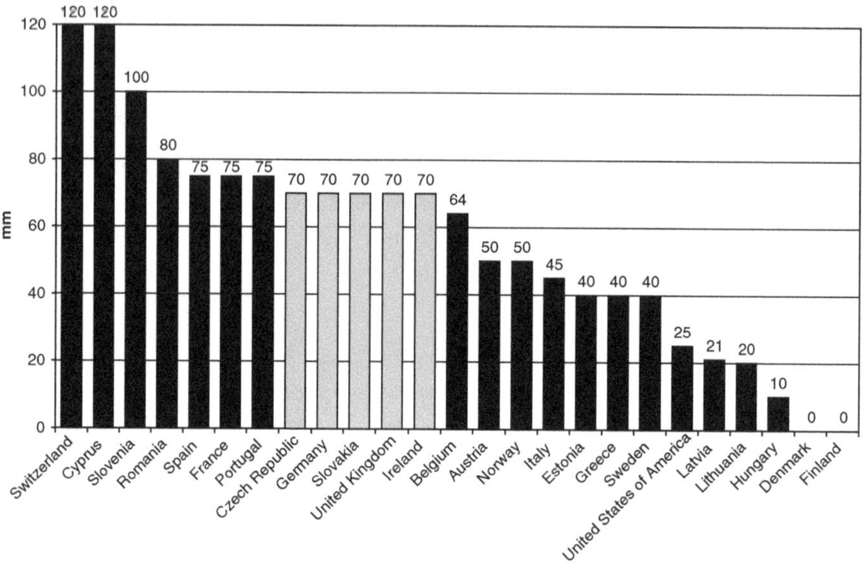

Fig. 3.5 Minimum dbh thresholds (mm)

Although dbh is among the ecological field variables for which the most complete and detailed information is available, harmonization of estimates requires a common minimum dbh threshold which for European countries is 120 cm. However, 23 of 26 countries use thresholds less than 80 mm. For purposes of carbon assessments under the Kyoto Protocol, a small standardized minimum threshold would be desirable.

Most countries do not use a minimum threshold for measuring tree heights, perhaps because use of a minimum dbh threshold lessens the need for a height threshold. Among the 11 countries with a height threshold, five different thresholds are used with 1.3 m the most common. Often, tree heights are measured only for small sub-samples of trees with heights for the remaining trees predicted using models.

3.3.3.3 Number of Trees per Unit Area

Data from fixed area circular plots as well as from Bitterlich relascope assessments (Bitterlich 1984) permit estimation of stem density in terms of number of trees per unit area. Almost all countries (96%) are able to provide this information at the plot level. Apart from the effects of different minimum dbh thresholds, stand density estimates that are independent of plot size and sampling method could be useful for assessing the spatial distribution and variation in stem attributes.

3.3.3.4 Social Position of Trees

The social position of trees is assessed by the NFIs of more than 80% of the participating countries. At the very least, the national classifications of all countries distinguish between dominant and suppressed trees (Table 3.1). The Kraft (1884) and IUFRO (Leibundgut 1956) classifications are the most common approaches for assessing the vitality and competitive intensity of trees. The Kraft classification defines five social classes, whereas the IUFRO classification uses nine vitality classes. The Kraft social classes describe the current position of a tree relative to height in a mostly even-aged forest with a homogenous structure. With the Kraft classes, predominant trees are taller than dominant trees, while co-dominant, dominated and suppressed trees are defined without regard to forest dynamics and life cycles. The IUFRO classification system is based on three components, each with a number code: the first component refers to one of three height classes; the second component refers to tree vitality; and the third component refers to the near future growth potential.

Denmark, Norway and Sweden assess social positions of sample trees; and Greece, Hungary, Lithuania, Portugal and the USA assess social position or layering without providing a detailed method. For purposes of assessing biodiversity, at least the proportions of dominant, intermediate and suppressed trees could be reported. Suppressed trees have lower vitality than dominant trees, are vulnerable to colonisation by insects and fungal species, and contribute to increasing forest biodiversity.

3.3.3.5 Forest Layering

Vertical forest structure is characterized by most NFIs in terms of crown layers in nominal categories (Table 3.2). All 25 countries that record layers include a category for plots with only a single layer. In the Wallon region of Belgium small plantations, young high forest (trees grown from seed) and coppice are included in

Table 3.1 Classifications describing social position and dominance

Number of classes	Classes	Number of countries
No assessment	–	5
2	Dominant, dominated dominant and co-dominant	3
3	Dominant, co-dominant, suppressed	2
4	Dominant, co-dominant, intermediate, suppressed	2
5	KRAFT predominant, dominant, co-dominant, dominated, suppressed	6
6	–	1
7	–	1
9	IUFRO classification describes vitality of trees in the upper layer, the middle layer, lowest layer	3

Table 3.2 Assessment of tree layers

	Number of layers		
Country	One layer	Two layers	Three- or multi-layered, mixed layered, uneven-aged
Austria	x	x	x
Belgium (Wallonia)	x	x (or more)	–
Cyprus	x	x	x
Czech Republic	x	x	x
Denmark	x	x	x
Estonia	x	x	x
Finland	x	x	x
France	"No defined layers" only two different vertical structures		
Germany	x	x	x
Hungary	x	x	x
Iceland	x	x	–
Ireland	x	x	x
Italy	x	x	–
Latvia	x	x	x
Lithuania	x	x	x
Norway	x	x	x
Portugal	Assessment of the cover (%) of the three predominant species in seven height classes		
Romania	x	x (or more)	–
Slovakia	x	x	x
Slovenia	x	–	x
Spain	x	x (or more)	–
Sweden	x	x	x
Switzerland	x	x	x
United Kingdom	x	x	x
USA	x	x	x

this category. The second most frequently used category is for plots with two layers; three countries extend this category to two or more layers. However, most countries use a category for plots that have three layers.

At a first level of analysis, country assessments of layering are similar in the sense that they aggregate tree heights into vertical stratifications. However, a second level of analysis must consider the layer definitions used by countries. Three aspects must be considered: (1) minimum threshold of the lowest layer, (2) height thresholds of the layers and (3) the minimum coverage to qualify as a layer. Minimum national height thresholds vary considerably from 0.1 m in Spain, 0.4 m in Switzerland, 1.3 m in Austria, to 4.0 m in Estonia. Consequently, a stand with an upper crown layer with 60% coverage and an understory of 1 m height with 20% coverage will be recorded in Estonia and Austria as single-layered and in Sweden and Switzerland as two-layered. Further, in Finland the lowest layer must provide sufficient regeneration for development of the next stand to be considered as a layer. Slovenia defines height layers using dominant tree dbh in six classes with a special emphasis on young trees. Although the classification is based on dbh, it is considered in Slovenia as a height classification because of the strong correlation between height and dbh.

Most countries do not define height ranges for the layer classes. However, Austria requires that the height difference of two layers must be one third of the average height of the tallest layer, a criterion that limits the number of layers to three. Spain uses a similar approach by requiring that differences in heights between layers must be one-third of the average height. In Estonia, the height of the second layer must be 25–75% of the height of the upper layer but at least 4 m. In Latvia trees with height differences of 20% of the average tree height are merged into the first layer, and trees with greater differences are assigned to the second layer; existing undergrowth trees are regarded as a third layer.

Minimum coverage is required and defined by four countries. Switzerland and Spain require a minimum of 20% crown cover for a layer, while Austria requires a 30% crown cover. Sweden requires that each layer must have at least 5 m^2 basal area per ha or a minimum of 500 stems per ha with dbh greater than 10 cm.

Because of lack of comparable data and quite different definitions of layers, harmonization of forest layering may be less feasible. However, many countries assess vertical layering, and it is an inexpensive field measurement. In addition, the variable is a nominal attribute that is recorded by visual assessment. Therefore, a minimal level of harmonization could be achieved by using an indicator assessing the proportions of one-, two- and multi-layered forests.

3.3.4 Harmonization Possibilities

Proposals for harmonizing estimates of indicators of tree species composition, horizontal structure, and vertical structure within classes of forest categories (Sect. 3.2) were tested using data from the common database (Chap. 4). For the tests, tree species composition was assessed within forest categories with respect to two features: (1) the relative abundance of native tree species expressed in terms of proportion of total plot basal area, and (2) the proportions of plots with 1, 2, 3, 4, or more native tree species. Horizontal forest structure was assessed within forest categories with respect to two features of dbh distributions: (1) mean dbh of the 0.1%, 1%, 5% and 10% of trees with largest dbhs. Respondents to the first WG3 questionnaire regarded the latter indicator of the maintenance of old growth characteristics as both important and feasible; and (2) mean of the plot-level standard deviation of dbh. Vertical forest structure was assessed within forest categories with respect to two features: (1) plot-level standard deviation of tree height; and (2) proportions of one- two- and multi-layered plots using current national definitions. The results of all analyses are reported in Sect. 5.3.

3.4 Forest Age

3.4.1 Background

The ages of trees and stands, as well as the age structures of forest, are typical and major features of forest ecosystems. Tree age is a good candidate for a biodiversity indicator because large and old trees are important habitats for typical forest animals such as black storks, woodpeckers, small mammals, bats, and beetles and for lichens, fungi and bryophytes (McGee and Kimmerer 2002; Rambo and Muir 1998a; Brändli et al. 2007a). Epiphytic, saprophytic and saproxylic species that grow or spread slowly or that follow each other in a succession on the same tree particularly depend on old trees and stands. Cavities in large trees filled with woody humus are rare but important habitats in natural forests, and their quantities are correlated with tree and stand age. Rare forest lichens often occupy the trunks of large living trees (Gilg 2005), and even forest ground species such as molluscs may have greater richness in older stands (Brändli et al. 2007a). Correlations between stand age and species density are known for vascular plants (Deconchat and Balent 2001; Pitkänen 1997; Halpern and Spies 1995), bryophytes and fungi (Jonsson and Jonsell 1999; Rambo and Muir 1998b), molluscs (Müller et al. 2005) and birds (Poulsen 2002). Because the ecological impact of old stands is greater than that of single old trees, ecologists postulate creating islands[2] of old growth forest interspersed in managed forests as a means of fostering connectivity among different habitat types (e.g. Scherzinger 1966).

The age structure of forest areas reflects the effects of natural and human disturbances with older forests usually indicating more natural dynamics. Older forests better represent the ageing and decay phase of natural forests with long life cycles such as 400 years in a natural spruce-fire-beech mountain forest. However, because of intensive forest management and harvesting, most European forests are relatively young. In even-aged forests, the percentage of stands older than 140 years is 2% in Central Europe and 4–5% in Eastern and Nordic/Baltic Europe (MCPFE 2007). Indicators for stand age based on proportions such as proportion of stands with trees older than a specified age are useful and relevant for assessing the biodiversity component of sustainable forest management. Thus, forest age has frequently been proposed as biodiversity indicator (Bosch and Söderbäck 1997; UNEP 2001; EEA 2003; Brändli et. al. 2007b).

3.4.2 Forest Age Assessment within NFIs

Although knowledge of age class distribution is a fundamental prerequisite for assessing the long-term sustainability of forest management interventions and wood supply, too little information regarding forest age can be found in international

[2] Stands of 0.5–5.0 ha of old trees with dbh $\geq$ 45 cm at distance of less than 1 km.

reports (MCPFE 2007). A primary problem is estimation of age for uneven-aged stands. Another primary problem is that information on age is not available for an increasing proportion of stands. The latter problem may be attributed to several factors. First, most countries only assess and report age for even-aged forests. Second, for a variety of reasons such as close-to-nature silvicultural practices, the proportion of uneven-aged stands has increased in recent decades. Third, definitions of the terms *even-aged* and *uneven-aged* are not available for most countries, and where they are available they are not comparable. Thus, the potential for constructing a bridge to harmonize estimates of age is relatively small. Instead, we propose a new indicator, dominant age, that can be applied to all kind of stands (Sect. 5.4).

Forest age data are available for the NFIs of 24 countries responding to the questionnaire (Fig. 3.6). Of these 24 countries, 18 assess tree age and 23 assess information on stand age or can calculate it using tree ages. Thus, 96% of the NFIs of responding European countries provide data on forest age which makes it an excellent candidate for a biodiversity indicator.

The NFIs of all responding countries use tree age definitions commonly included in forestry textbooks. Tree age is defined as biological or actual age of individual trees determined as the time elapsed between the germination and measurement dates. Tree age in coppice systems is defined as the biological age of the above-ground tree stem, not the age of the rootstock or the total age from seed. In a plantation forest, age is based on the planting date.

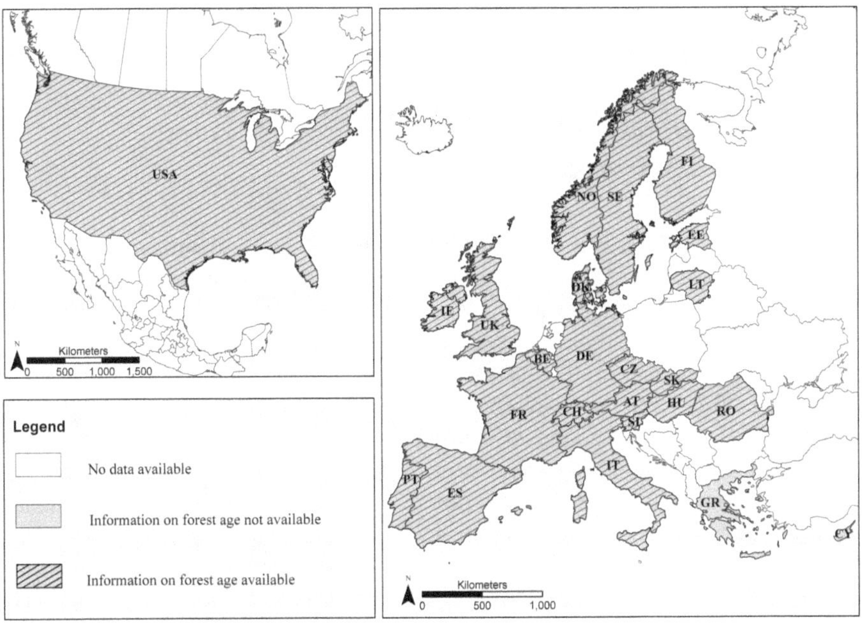

Fig. 3.6 Countries whose NFIs assess forest age

Table 3.3 Generalized country definitions of stand age

Generalized definition	Number of countries
Mean age of the trees in the upper (dominant) tree layer (main stand); mean age of (co-) dominant trees (from the over story)	8
Mean tree age of the dominant species (in the upper layer)	3
Mean age of the trees making up 80% of the growing stock	1
Mean age of the most important stand element(s) regarding management objectives	2
Mean age of all trees	2
Mean age of all trees weighted with basal area or crown cover or storey cover	3
Median of tree ages	1
Time between years of planting and inventory	3

Definitions of stand age vary widely with respect to which trees are considered and how tree age is defined (Table 3.3). Traditionally, stand age has been used for economic or silvicultural purposes. Most stand age definitions are based on the dominant stand elements such as the dominant tree species, the upper or dominant layer, or trees constituting 80% of growing stock. However, some definitions are based on all trees in the stand, all the trees satisfying the minimum dbh threshold, or all trees belonging to the dominant species. The NFIs of only a few countries estimate stand age as a weighted average or the median of the ages of all living trees in a stand. For plantations, some countries estimate stand age as time elapsed since the planting year; others use age at breast height; but most use biological age.

3.4.3 *Similarities and Differences in the NFIs Forest Age Assessment*

All European countries use similar definitions of tree age and similar methods for assessing tree age. Basically, tree age is assessed using two methods: (1) tree ring analysis and (2) other estimation methods such as counting annual shoots/branch/whorls, local interviews, visual assessments, counting annual rings on stumps, and forest management records. Frequently, combinations of the above methods are used.

Among NFIs, different assessment methods and approaches should provide comparable data. However, the NFIs of most countries base stand age assessments on ages of only a sample of stand trees. Usually the representative trees are not chosen randomly but rather represent dominant species, the main layer or dominant diameter. Only five countries assess the age of all sample trees on a NFI plot. Thus, 80% of countries cannot provide tree age data for the entire population. An intermediate conclusion is that existing tree age data for European NFIs do not permit direct comparative analyses. The solution might be to make comparisons using stand age.

Definitions of the terms *stand*, *stand age*, *even-aged* and *uneven-aged* used by NFIs are similar but are not harmonized in most important details. As used by European NFIs, stand age entails two concepts or definitions: mean age of dominant trees which is used by most countries, or mean age of all trees which is used by approximately 20% of countries. Ecologically, the former concept is better related to biodiversity, is much less dependent on the measurement threshold and corresponds to the majority of definitions in the literature (e.g. Helms 1998). Plantation age is generally based on the year the plantation was established without regard to the age of the nursery stock.

In general, stand age is assessed using two main methods: direct field assessment or indirect assessment by calculations based on sample tree ages. Stand age is assessed in years by the NFIs of all countries except three which assess stand age by classes. The NFIs of most countries differ in the methods used to assess tree and stand age. Some countries use the very objective method of drilling trees almost exclusively, whereas other countries use simple estimation methods. When using tree ring analyses with permanent plots, trees growing outside the plots are usually drilled. The number of trees used to assess tree age for stand age estimation varies from one tree to all plot trees. Stand age assessment is generally related to trees on the plot (plot age) but rarely to larger stands as in Hungary or Romania. Plot sizes vary from 100 to 5,000 m^2 (Table 3.4). Although only half the countries have similarly sized reference areas (e.g., 500–1,000 m^2), stand age assessed at the plot level is in general comparable from a methodical point of view. Nevertheless, the objectivity and reproducibility of stand age assessments differ considerably.

Many countries do not assess age in uneven-aged stands and coppice with standards.[3] Thus, the reference forest area for which age estimates are available is only 50–60% of the total forest area in some countries. To compare proportions of old stands (e.g., older than 160 years) country-by-country would be difficult based on currently available stand age information. Further, for countries that assess stand age in classes, the classes have not been harmonized and the age boundary for the oldest class varies from 60 years to 120 years to 140 years.

A reasonable conclusion is that existing forest age data for European NFIs permits only restricted analyses at the international level; e.g., the proportion of stands older than 120 years in even-aged high forests. However, because the definitions and proportions of even-aged stands differ greatly by country, the utility of age as a biodiversity indicator is doubtful. Comparisons of stand age may be possible in the future if countries agree to assess age for all stands. Spain and Switzerland have already proposed to do so as the result of Cost Action E43 investigations and results. Countries that already assess age in uneven-aged stands focus primarily on the dominant trees of the stand or plot which are generally represented by the largest trees. From an ecological perspective, dominant age, Age_{dom}, is proposed as a new indicator that can used for all kind of stands.

[3] Standards are trees that generated from seed and that are left in harvested areas to support sustainable production of both timber and non-timber products.

Table 3.4 Reference area for stand age assessment and forest area with age assessment

Country	Reference area for stand age assessment (plot or stand size): Minimal	Maximal	Forest area with stand age assessment: Even-aged high forest	Coppice forest	Coppice with standards	Uneven-aged stands	Other wooded land
Austria	500 m²	–	X	X	X	X	–
Belgium	1,018 m²	1,018 m²	X	–	–	–	–
Denmark	1,000 m²	–	X	X	–	–	–
Estonia	500 m²	–	X	X	–	X	–
Finland	2,500 m²	5,000 m²	X	Not practiced in FI		X	X
France	707 m²	707 m²	X	X	X	–	–
Germany	Variable plot size		X	Calculation possible		X	–
Hungary	<1,000 m²	>150,000 m²	X	X	X	X	–
Italy	530 m²	530 m²	X	X	X	–	–
Lithuania	1,000 m²	–	X	X	X	X	–
Norway	1,000 m²	1,000 m²	X	–	–	X	–
Portugal	250 m²	2,000 m²	X	X	–	–	–
Romania	5,000 m²	200,000 m²	X	X	Not applied	X	X
Slovakia	100 m²	500 m²	X	X	X	X	X
Slovenia	500 m²	–	X	X	–	–	–
Spain	2,000 m²	2,000 m²	X	X	–	–	–
Sweden	1,256 m²	1,256 m²	X	X	X	X	X
Switzerland	500 m²	2,500 m²	X	X	–	–	–
UK	500 m²	–	X	X	X	–	–

3.4.4 *Harmonization Possibilities*

To assess the necessity for reference definitions and harmonization of estimates and the possibilities for comparing estimates of forest age, the harmonization test consisted of using the common database to address the following questions (Chap. 4):

1. What is the influence of different national definitions on the comparability of stand age estimates?
2. To what degree do estimates of plot ages depend on the dbh measurement threshold?
3. What is the correlation between tree age and dbh?
4. Can bridges in the form of models be constructed to predict tree age from other variables?
5. Are the proposed indicators practicable and do they produce comparable results?

The results of the tests are reported in Sect. 5.4.

3.5 Deadwood

3.5.1 Background

Standing dead trees, dead branches and fallen logs collectively constitute one of the most important European forest habitats for wildlife. As many as one-third of European forest species depend on deadwood for their survival (Boddy 2001; Siitonen 2001). Deadwood provides habitat, shelter and food for birds, bats and other mammals and is particularly important for the less visible majority of forest species including insects, beetles, fungi, bryophytes and lichens. Deadwood also has a key role in sustaining forest productivity, forest equilibrium and resilience and carbon sequestration. Despite its importance, deadwood is now at critically low levels in many European countries, primarily because of management practices used in commercial forests and even in protected areas. Typical European forests have less than 5% of the deadwood expected for natural conditions (WWF 2004). For many European and international agreements, deadwood is increasingly selected as a key indicator of forest naturalness and sustainable forest management (MCPFE 2003). Forest inventories are crucial for evaluation of the state of deadwood in European forests. Although deadwood was generally not assessed before the 1990s, many NFIs have since added it as a core variable. The NFIs of Finland, Norway, Sweden and Switzerland were among the first to assess deadwood (Stokland et al. 2004; Böhl and Brändli 2007).

In recent years, the NFIs of many countries in Europe and North America have begun to monitor deadwood and have become potential sources of reliable, comparable information on deadwood at national scales. However, techniques for cross-country harmonization of deadwood estimates are generally not available. To gain a better overview of deadwood assessments undertaken in NFIs and the potential for harmonization, two questionnaires were distributed by WG3 to participating countries. Responses to the questionnaires, together with new data coming from countries having joined the Action more recently, have produced the results presented in this section.

3.5.2 Deadwood Assessment Within NFIs

Twenty-three European countries and the USA responded with information on deadwood (Fig. 3.7). Plot areas for assessing deadwood range from 154 to 706 m^2 (Table 3.5). The NFIs of ten countries use a single circular plot; the NFIs of two countries use line intersect sampling; one NFI uses three 10-m transects overlaid on a 500 m^2 circular sample plot; and one NFI uses two 25×4 m^2 transects. The NFIs of four countries use variable radius plots for sampling deadwood: for three, plot sizes are based on sizes of deadwood pieces; for the other, plot size depends on the mean size of living trees.

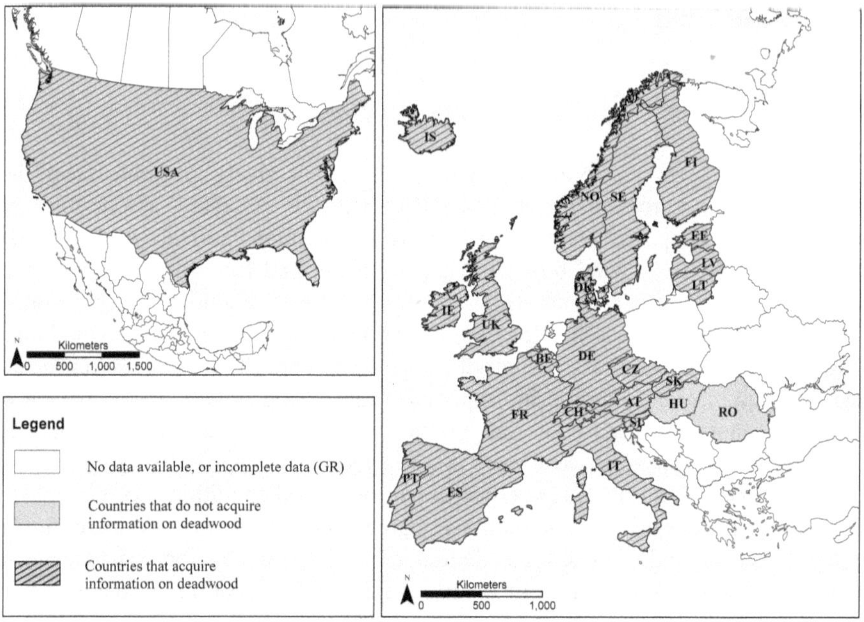

Fig. 3.7 Countries whose NFIs assess deadwood

Table 3.5 Number of plots and plot area by country

Country	Number of deadwood plots	Deadwood plot area[a] (m²)
Austria	10,000	300
Belgium	10,800	VAR
Czech Republic	14,500	500
Estonia	1,000	314
Spain	–	706.86
Finland	66,000	154
France	70,000	700
Germany	54,000	78.5
Italy	7,000	530
Latvia	10,000	500
Lithuania	5,600	VAR
Norway	10,000	250
Slovakia	16,000	LIS
Sweden	4,300	VAR
Switzerland	6,500	LIS
United Kingdom	–	VAR
USA	34,000	LIS

[a] *VAR* Variable Radius Sampling, *LIS* Line Intersect Sampling

3.5.3 Similarities and Differences in the NFIs Deadwood Assessment

Generally, larger deadwood components are more likely to be assessed by NFIs (Table 3.6). Uprooted stems (windthrown trees) are measured by all the responding countries that assess deadwood, whereas fine woody debris is less frequently assessed, perhaps because of the time necessary to do so. The variety of deadwood definitions impedes calculation of comparable estimates among NFIs. The NFIs of at least three countries include stumps when assessing deadwood, one NFI includes volumes of dead limbs with diameters less than 5 cm, and one NFI includes only dead trees that have died within the last 5 years. Because most NFIs do not distinguish among a large number of deadwood categories, and to simplify the summary of questionnaire responses, WG3 aggregated deadwood components into three categories: (1) intact and broken snags, (2) uprooted trees and stems, and (3) lying deadwood which includes clear cut stems, pieces of stems, pieces of branches, cut branches, logging residues, fine woody debris, and broken and lying stems without uprooting.

The NFIs of all countries use minimum deadwood dimension thresholds for sampling purposes, because assessment of all deadwood pieces is not possible (Table 3.7). However, the thresholds vary considerably. For the Czech Republic, intact snags are considered stumps, and uprooted staves, defined as residue after cutting uprooted stems, with diameters greater than 30 cm are measured. For the Slovak Republic, logging residues and fine woody debris with diameters greater than 1 cm are measured. For France, clearcut stems are considered deadwood. For Slovenia, a minimum diameter threshold of 10 cm is used for some plots, while a minimum diameter of 30 cm is used for other plots. For Lithuania, only standing and lying dead trees suitable for firewood are measured.

The exact positions at which deadwood diameter measurements are made varies considerably among NFIs: it may be in the middle of the piece, at the butt end, at the thin end, or for line intercept sampling (LIS), it may be at the point where the

Table 3.6 Countries including deadwood components in their definitions

Components	Number of countries
Uprooted stems	23
Clearcut stems	17
Pieces of stems	21
Pieces of branches	16
Cut branches	15
Uprooted staves	12
Logging residues	17
Fine woody debris	12
Intact snags	23
Broken snags	23
Broken, lying stems without uprooting	22

Table 3.7 Number of countries using minimum diameter thresholds by deadwood types

	Number of countries		
	Deadwood types		
Minimum diameter (cm)[a]	Snags	Uprooted trees and stems	Lying deadwood
4	2	1	1
5	3	3	3
6.1	1	1	1
6.4	1	1	1
7	3	4	4
7.5	3	2	2
7.6	–	1	1
10	5	6	6
12	1	–	–
12.7	1	–	–
15	2	1	1
20	1	2	2

[a]Mean minimum diameter: 9.4 cm

Table 3.8 Number of decay classes

Number of classes (decay stages)	Number of countries
2	2
3	5
4	7
5	6
6	1

piece intercects the sample line. Because diameter measurements are often crucial in determining if a deadwood piece is to be included in the sample, these different measurement positions may produce quite different results.

Among the NFIs that assess deadwood, 17 characterize the level of decomposition of deadwood pieces and trees. The number of levels or stages of decay varies among countries (Table 3.8). The mean number of levels is four and could be considered a good basis for a reference definition.

Most NFIs use two different ways to calculate dead wood volume: volume tables for standing and lying dead trees, and Huber or Smalian methods (Rondeux 1999; Avery and Burkhardt 2002) for lying dead wood. The volume of standing snags may be calculated with tables or with Huber or Smalian methods. Among the NFIs responding to the questionnaire, one does not estimate deadwood volume, and one uses several methods depending on species and deadwood category (Table 3.9).

The responses to the questionnaire highlight the great variation in methods used by NFIs to assess deadwood. This variation is evident in many aspects of deadwood assessments including sampling methods, types and sizes of plots, dimensional thresholds, attributes, and volume estimation methods. The selection of a sampling

Table 3.9 Deadwood volume estimation methods

Method	Number of countries	
	Standing deadwood (snags)	Lying deadwood
Volume models based on dbh or dbh and height	9	2
Huber	4	8
Smalian	2	2
Line intersect sampling	–	2

method depends on practical contingencies such as time availability, financial constraints, and NFI objectives. Variation in thresholds is among the most important items that must be considered when constructing bridges for harmonization. For other items such as variation in decay classes compromises may be found by regrouping classes.

3.5.4 Harmonization Possibilities

Deadwood assessments by NFIs in Europe and the USA are based on a variety of locally adapted definitions, algorithms, thresholds, categories and indicators. Because the NFIs of many countries assess deadwood, development of methods contributing to harmonized estimation and reporting merits serious consideration (Rondeux and Sanchez 2009). To facilitate this effort a consistent set of deadwood reference definitions is needed.

The main deadwood variable reported by all the NFIs is total volume with additional information such as the species of the trees to which the deadwood elements belong, the spatial position (lying or standing) and the decomposition stage. Deadwood assessments conducted by NFIs generally consist of four steps (Chirici et al. submitted):

Step 1: Deadwood pieces are included in samples if they simultaneously satisfy the definition of a deadwood component and sampling requirements such as lying partially or entirely within plots for fixed area sampling or across sample lines for LIS.

Step 2: The characteristics of deadwood pieces such as dimensions, tree species and decay stage are assessed in the field using protocols consistent with the sampling methods described in Step 1.

Step 3: Piece-level attributes such as volume are estimated using the Step 2 measurements.

Step 4: Plot-level estimates are aggregated to produce large area or population estimates using appropriate statistical estimators.

The second WG3 questionnaire inquired regarding methods used by NFIs to assess deadwood with emphasis on Steps 1, 2 and part of 3 above. One finding was that

NFIs use a variety of protocols and definitions in these steps. In the first step, three main differences affect the potential for harmonizing deadwood estimates:

1. Definitions that distinguish between living and deadwood;
2. Definitions of deadwood components such as snags or stumps and protocols for characterizing a deadwood piece as standing or lying and for assigning it to the appropriate component;
3. Dimensional thresholds for definitions such as minimum diameter and length for lying deadwood, minimum dbh and height for standing deadwood, minimum diameter and/or height for stumps; minimum diameter thresholds for both lying and standing deadwood vary between 4 and 20 cm and length/height thresholds vary between 0 and 200 cm.

Other first-step differences such as sampling method (plot or line), number of sample units, relative locations within sample units, and field measurement protocols should have a negligible impact on the ability to harmonize large area or population estimates produced by different NFIs. However, differences in second-step field measurement protocols were also evident:

a. Deadwood volume and biomass: biomass is generally estimated from volume, and volume is based on dimensional measurements which, in turn, depend on definitions and thresholds such as minimum diameter and length for lying deadwood pieces, minimum height for standing deadwood, whether volume is estimated over- or under-bark, and whether small branches or stem top are included or excluded.
b. Decomposition stage: this information, which is particularly relevant for assessing the ecological value of deadwood, is assessed with very different systems of nomenclature and with numbers of classes ranging between two and seven; however, even if the number of classes is similar, class definitions are frequently based on different underlying classification criteria for the apparent characteristics of decomposing wood such as softness, texture, or colour.
c. Species or species group: because deadwood species identification is difficult in the field, NFIs typically just use broad classes such as broadleaf and coniferous.

Differences in third-step procedures, such as characterizing pieces of lying deadwood as cylinders or truncated cones, affect volume estimates for individual deadwood pieces.

In general, the effects of differences in these methods on final deadwood volume estimates are either known or can be determined because the mathematical relationships among the different models are already documented or can be empirically investigated (Rondeux et al. submitted).

This is the topic of the harmonization test whose results are reported in Sect. 5.5. The test was structured using four steps: (1) construction of international reference definitions, (2) construction of bridges that convert estimates based on national definitions to estimates based on the reference definitions, (3) application of these bridges to raw NFI data, and (4) evaluation of harmonized estimates.

3.6 Regeneration

3.6.1 Background

Forest regeneration is recognized as an indicator of sustainable forest management. It has also been specified as an indicator of forest biodiversity within the guidelines of sustainable forestry issued through the Ministerial Conferences on the Protection of Forests in Europe (MCPFE 2007) and the Montréal Process (2006). Environmental change due to factors such as climate change, invasive species, habitat fragmentation and eutrophication of forest soils is rife and poses real threats to the regeneration potential of numerous tree species within their historical natural distributional ranges. Thus, recognition of forest regeneration as a key indicator is crucial. Surveillance of forest recruitment success or failure is therefore of increasing importance to provide forward-looking guidelines for forest managers and conservationists.

Where monitoring reveals regeneration failure, predictions can be made regarding the likely development of the stand including likely changes to tree species composition, nutrient and energy dynamics, hydrological and microclimatic conditions, forest and food web structure and forest biodiversity (Ellison et al. 2005). For example, in the hardwood forests of eastern North America monitoring of forest regeneration has revealed widespread regeneration failure of oak, a condition thought to be related to climatic variables (Aldrich et al. 2005). The white bark pine (*Pinus albicaulis*), common in the Rocky Mountains as a dominant, late succession tree species, is similarly suffering regeneration failure, although in this case the cause is fire suppression. A similar wholesale change in forest ecosystem structure and functioning has evolved in floodplain woodlands worldwide where river management activities have altered the natural periodicity and magnitude of flood disturbances with substantial disruptions to the recruitment success of Salicaceae species (*Salix*, *Populus*) (Barsoum 2002). In all these cases, the consequences of a loss in the dominant tree species results in a loss of dependent species and a simplification and homogenisation of forest ecosystem structure.

Trends observed in natural forest regeneration may prompt forest managers to take any of several actions. First natural regeneration may be promoted by the reintroduction of natural disturbances such as fires and periodic flooding and/or by controlling invasive species that stifle recruitment. For example, *Rhododendron ponticum* is a non-indigenous invasive shrub species in the Caledonian pine and Atlantic oakwood forests of Scotland that prevents natural tree species regeneration through competitive exclusion (Palmer et al. 2004). Second, artificial forest regeneration methods such as direct seeding, planting of nursery-reared saplings and/or coppicing may be introduced. Where forest regeneration is primarily by artificial regeneration strategies such as in Finland, Sweden and the United Kingdom (Hummel 1991; Pukkala 2006; Axelsson et al. 2007), regeneration monitoring provides forest managers with a means of surveying the progress of a new crop, including the stages at which forest management interventions might be needed for purposes of controlling competing ground vegetation or protecting recruits from browsing mammals (Willoughby et al. 2004).

3.6.2 Regeneration Assessments within NFIs

Regeneration can originate through natural forms of recruitment either from seed or asexual regeneration strategies. This latter regeneration strategy can include sprouting from stems, roots, or broken branches and for certain tree species can be a dominant regeneration strategy (Tardif and Bergeron 1999; Barsoum et al. 2004; Douhovnikoff et al. 2004). Regeneration can also occur through artificial means of propagation such as direct seeding, nursery-raised saplings, or coppice re-growth. Effective regeneration monitoring relies on repeated surveys and focuses on new recruits that have passed the most vulnerable early recruitment phase, i.e. the focus is on saplings, rather than seedlings or early stages of vegetative sprouting. Many countries in Europe and North America actively monitor forest regeneration within their NFIs, and these are the most reliable sources of repeated assessments of forest regeneration at a regional scale.

The WG3 regeneration questionnaires were designed to provide an overview of where, how and over what time period regeneration has been monitored; responses were received from 25 countries including the USA. Respondents were requested to describe how their respective surveys defined regeneration and whether distinctions were made among different types of regeneration (from seed, vegetative, artificial) and among different vegetation species occurring within sample plots. Respondents were also requested to indicate whether the genetic origin of artificial recruitment via planting or direct seeding was recorded. Additionally, the respondents were requested to indicate the locations (e.g., clear-felled areas, under-canopy) where forest regeneration was sampled and whether disturbance variables known to influence regeneration were quantified. The questionnaire was divided into three main sections: (1) the sampling target which focused on vegetation types, vegetation species, types of regeneration, and genetic origin; (2) regeneration sampling methods which focused on number and configuration of subplots, assessment methods, sampling frequency, and permanence of regeneration subplots; and (3) characteristics of the sampling area including canopy cover, forest structure, and disturbances.

3.6.3 Availability of Information

As the survey aspect of this work progressed, it became apparent that the size range of recruits considered to be regeneration differed considerably among NFIs. For some NFIs, the smallest size class of saplings in tree volume assessments is considered regeneration for other NFIs. To maximize the opportunities for comparison of regeneration sampling populations and methodologies among NFIs, and thus, the scope for harmonization, the smallest size class of saplings in NFI tree volume assessments is included here as regeneration despite the fact that this size class may not specifically be included as part of an NFI's regeneration survey.

Of the 25 European countries and the USA that responded to the regeneration questionnaire, 25 assess regeneration in their NFIs (Fig. 3.8).

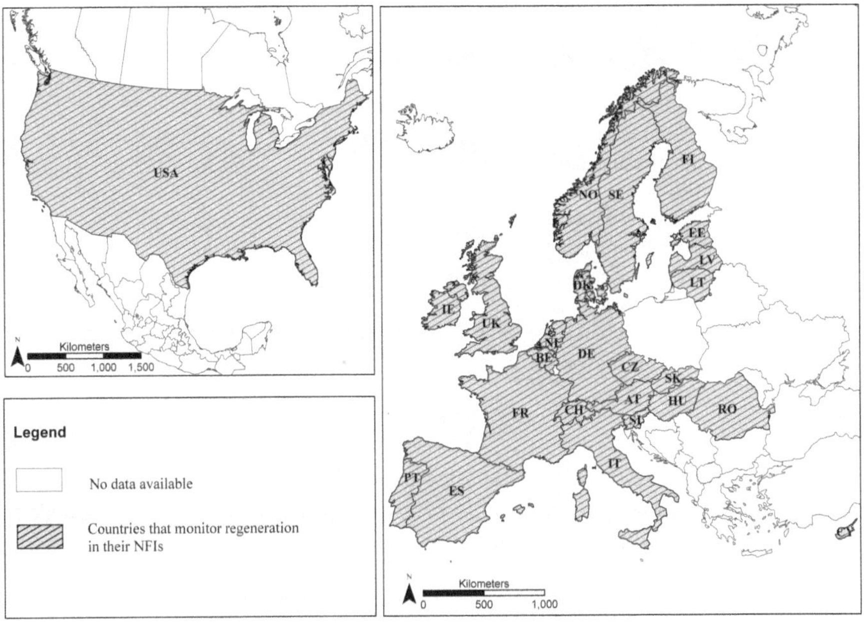

Fig. 3.8 European countries whose NFIs assess forest regeneration

NFIs that collect information on regeneration mostly do so for tree species only, although a small proportion of the responding countries also assess recruitment of shrub species (Table 3.10). None of the responding countries include the regeneration of non-woody ground vegetation species in their regeneration assessments. While the majority of countries record and distinguish among all regenerating tree species on a sample plot, one country simply combines all species and records regeneration as present or absent for a sample plot, and three countries only record regeneration for the dominant tree species.

Natural regeneration and certain forms of artificial regeneration are recognized as forms of regeneration by most NFIs (Table 3.10), regardless of predominant silvicultural practices. Distinctions do not always tend to be made, however, between the relative proportion of natural regeneration from seed versus vegetative regeneration strategies, or between the different types of artificial regeneration, although most NFIs distinguish between natural and artificial regeneration.

Coppice re-growth following stand management activities is frequently (60% of NFIs) recorded as a form of artificial regeneration in NFI surveys, although in some countries it is not, despite high frequencies of coppicing (UN/ECE-FAO, 2000). Where natural regeneration is assessed in NFIs, attempts to distinguish between regeneration from seed and regeneration from asexual regeneration strategies are made in only 36% of these NFIs (Table 3.10). The relative proportion of sexual versus vegetative regeneration is, therefore, unknown in the majority of cases. Similarly, with regard to artificial regeneration, few countries make a point of

Table 3.10 Sampling target for regeneration assessments in NFI surveys

Sample Target (Regeneration type)	Proportion of countries
Natural regeneration – from seed	1.00
Natural regeneration – asexual; i.e. re-sprouting from stems, roots but not from coppice	0.88
Distinction between sexual and asexual forms of natural regeneration	0.36
Artificial regeneration – direct seeding	0.92
Artificial regeneration – planting of saplings	1.00
Artificial Regeneration – coppice re-growth from roots/stems following stem removal	0.60
Distinction between different types of artificial regeneration	0.36
Distinction between genetic origin of artificial regeneration	0.04

distinguishing among the different types of artificial regeneration, and only one country attempts to record the genetic origin, or provenance of direct seeded, or planted recruits found in NFI regeneration survey plots.

3.6.4 Similarities and Differences in Regeneration Assessments

Most responding countries use at least one subplot of specified dimensions to record regeneration; only one country does not use subplots and instead makes estimates of forest regeneration coverage without regard to a predefined area (Table 3.11). Eleven of the NFIs do not have any replication of regeneration sampling plots, using only one subplot for each NFI survey point. The remainder of NFIs have between two and five regeneration subplots per NFI survey point and, exceptionally, Estonia uses 32 replicate regeneration subplots. In 96% of cases, subplots are circular; only two countries use square or rectangular subplots for a given regeneration sample population. In some NFIs, the number of subplots is modified according to the type of regeneration being assessed and/or the recruitment stage. Subplot dimensions also vary among NFIs and sometimes, depending on the regeneration sampling target, within an NFI survey (Table 3.11). Generally, where plots of different sizes are used, they tend to each be of a fixed dimension to fit a predefined regeneration sampling population. In some NFIs, however, the plot sizes are not fixed but rather depend on recruitment density or the size of the main NFI plot which may itself be of variable dimensions.

The majority of NFIs use a minimum and maximum height and/or dbh range to distinguish a new recruit of a given tree or shrub species (Table 3.11). Among the responding countries that have predefined size ranges for new recruits, many also group recruits into two or more size classes to distinguish the relative proportions at different stages of recruitment. These size classes are based only on height for seven NFIs, only on dbh for four NFIs and on both height and dbh for seven NFIs. Additionally, 12 NFIs have 2–3 different NFI assessment surveys that contribute

Table 3.11 Forest regeneration assessment methods

Regeneration sampling methodology	Proportion of countries
Subplots	
One regeneration subplot within NFI plot	0.50
Multiple regeneration subplots within NFI plot	0.50
Shape of subplot: Circular	0.92
Fixed size for predefined regeneration sample population(s)	0.92
Variable size – not dependant on predefined regeneration sample population(s)	0.13
Defining parameters for recruits	
Height (max/min size)	0.71
dbh (max/min size)	0.79
Different size/age classes using height	0.52
Different size/age classes using dbh	0.68
Assessment methods	
Presence/absence in subplot	0.04
Counts: Number in subplot	0.75
Counts: Number in different size classes in subplot	0.63
Plot area coverage (%)	0.29
Specific measurements of recruits in subplot (height, dbh, age, browsing damage, other)	0.21

information on forest regeneration. Each of these surveys potentially has its own selection of class sizes, target regeneration sample population and/or sample plot sizes. Some of these additional surveys, such as measurements of saplings as a component of tree volume assessments, are not known as regeneration surveys, but they yield further relevant information on forest recruitment to complement information obtained from the pre-specified regeneration subplots.

The majority of NFIs (75%) collect counts of the number of recruits in sample subplots (Table 3.11). A smaller fraction of these (63%), collect additional information on the number of recruits within different size classes in sample subplots. Other regeneration assessment methods used are (i) estimates of plot area coverage by recruits (29%), and in two cases, this is in addition to counts of the number of recruits in a subplot, and (ii) assessments of only the presence or absence of regeneration; only one NFI relies entirely on this latter method of assessment. A number of NFIs include specific measurements of individual recruits; three countries record height, four countries record dbh, one country records age, one country records the position of recruits in the plot, and one country records browsing damage.

In 54% of cases, regeneration assessments are undertaken in subplots situated within permanent NFI sample plots such that there is the possibility for subplots to be revisited at regular intervals. In 29% of NFIs, regeneration assessments are monitored in a combination of temporary and permanent sample plots. The remaining 16% collect information on regeneration in subplots situated solely within temporary NFI plots.

Regeneration has been monitored in many of the longer established NFIs for several decades, with other NFIs incorporating assessments of regeneration more recently between 2000 and 2007. None of the NFIs restricts sampling of regeneration to specific months of the year, although field work tends to take place in most NFIs only during the growing season, generally April to October in Europe and the USA.

In all NFIs, forest regeneration is assessed beneath the canopy and in 96% of NFIs also in open areas within the forested landscape. Further, in 71% of NFIs, efforts are made to collect records of disturbances that can affect regeneration success. In half of these NFIs, disturbance is simply registered as present or absent, while in the remaining cases a quantitative estimate of the proportion of the sample plots affected by a particular disturbance is given. In 40% of NFIs the timing of one or more disturbances is also recorded.

3.6.5 Harmonization Possibilities

Specific information might be missing in some NFIs regarding the types of forest regeneration under consideration; e.g. regeneration from seed, or vegetative fragments, type of artificial regeneration. However, there is no ambiguity over definitions of tree species regeneration and thus, the basic target sample population. There are difficulties, however, in making direct comparisons among regeneration datasets of different NFIs for which the differences in NFI sampling methodologies are considered. While the majority of NFIs count the number of tree species recruits in a given sample plot area, these counts are generally not directly comparable under three conditions: (i) the sample plot areas are not the same among NFIs, (ii) the frequency and timing of sampling intervals are not the same and (iii) the regeneration sample populations in NFIs are not defined in the same way in terms of their pre-defined size ranges as measured in terms of height and/or dbh. Differences in regeneration subplot areas and the numbers of replicate subplots among NFIs also suggest a strong likelihood of significant differences in sampling accuracy among NFIs. At present, comparisons among NFI forest regeneration data sets are, therefore, confined to assessments of proportional differences expressed as percentage of change in regeneration success from one stand to the next or from one sampling interval to the next within NFI data sets. Further, assumptions must be made regarding the consistency of forest regeneration assessment sampling methodologies among forest stands and among different sampling intervals within each NFI.

Conveniently, for the purposes of harmonization, there is consistency among many NFIs in the seedling/sapling height upper limit (130 cm) used to decide on the inclusion or exclusion of a recruit. This single common assessment parameter may offer a starting point for the harmonization of regeneration assessments in NFIs.

3.7 Ground Vegetation

3.7.1 Background

Quantification of biological diversity is an important objective in the assessment of non-timber resources in forest surveys (Groombridge and Jenkins 1996). Because the majority of forest plant diversity occurs in the ground layer in the temperate zone, sampling this community is particularly important (Johnson et al. 2006).

Ground vegetation assessment techniques rely on estimation of numerous well-accepted biodiversity indicators. For example, threatened species is an indicator selected by both sustainable forestry agreements, the MCPFE (1998) and the Montréal Process (1998). Differentiation between indigenous and introduced species is an indicator defined by the Lisbon Ministerial Conference (MCPFE 1998) in the context of production forestry. Structural indicators related to the assessment of ground vegetation as forest biomass were identified in the fifth Ministerial Conference (MCPFE 2007).

Development of a common definition for ground vegetation is difficult, primarily because ground vegetation is a vague expression that includes multiple life-forms (Alberdi et al. 2010). Major life-forms are usually designated by *tree*, *shrub*, *grasses* (Bonham 1989), although *ferns* and *bryophytes* are also sometimes monitored. Therefore, depending on the plant ecology requirements and the objectives of each NFI, different classifications provide details on different components of ground vegetation (Alberdi et al. 2010) (Table 3.12).

Ground vegetation relates to both forest types and forest structure (Pitkänen 1998) with each forest type having a specific associated understory. Tree species modify site conditions, soil chemistry, litter coverage, and light penetration, and these modifications lead to modification of ground vegetation (Augusto et al. 2003; Gärtner and Reif 2005). In addition, ground vegetation is highly correlated with fertility and stand age (Pitkänen 1997) and regeneration (Baier et al. 2005, Dusan et al. 2007).

Single species or species groups of forest ground vegetation can be used as indicators for site conditions (Khanina et al. 2007; Wilson et al. 2001), potential productivity, economic value, wildlife forage and shelter. Changes in composition and spatial arrangement of vascular plants may indicate the presence of chronic stresses such as site degradation (COST E43 2005). Ground vegetation has been also used to detect changes in the ecosystem due to air pollution, particularly nitrogen deposition, and climate change. Vegetation studies have the advantage of low cost relative to analyses of air or soil chemistry (Thimonier et al. 2003).

Data on the diversity of plant communities also provide information regarding the structure and productive base for all other organisms. This phenomenon indicates the potential utility of ground vegetation as a broad indicator of biodiversity because sampling plant communities is more efficient than extensive inventories of all biota (Johnson et al. 2006). Selection of a plot configuration and size for monitoring ground vegetation life-forms is a crucial part of the inventory design. Guidance on plot sizes for temperate ecosystems is reported in the literature

Table 3.12 Definitions and thresholds for ground vegetation layers

	Small trees	Shrubs	Herbaceous	Bryophytes	Lichens
FRA 2005 (FAO 2005)	–	0.5–5 m	–	–	–
BIOFOREST (Smith et al. 2005)	–	Large shrub layer: 2–5 m	Graminoid layer	Mosses and liverworts	–
		Subshrub layer: 0.5–2 m	Forb layer		
FOREST BIOTA (Granke 2006)	–	>0.5 m	Non-ligneous and ligneous <0.5 m	Moss layer (terricolous bryophytes and lichens)	Moss layer (terricolous bryophytes and lichens)
FLORIDA (Terminology from University of Florida)	–	Not height threshold defined	Herbaceous vegetation (non-ligneous)	–	–
EMAN (Roberts-Pichette and Gillespie 1999)	4 cm < dbh < 10 cm	<4 cm dbh	Non-ligneous and ligneous < 1 m	Ground layer vege tation (mosses, lichens, fungi and small trailing and rosette plants)	Ground layer vegetation
BIOCONDITION (Eyre et al. 2006)	–	Multi-stemmed or <2 m	Ground cover: grasses, herbs and forbs	–	–
			Weed cover		

(Aguilo et al. 1992) (Table 3.13) including recommendations of less than 0.1 m^2 for bryophytes, lichens and higher plant individuals and up to 200 m^2 for recording entire plant communities.

It is remarkable that in the frame work of the European Network of the International Cooperative Programme–Forest (ICP) (2009), the expert panel on ground vegetation assessment (Aamlid et al. 2007) defined a minimal area of 400 m^2, which is twice the previously recommended area, to achieve comparability of results among countries.

Because ground vegetation varies both seasonally and temporally, NFI cycles and measurement seasons are extremely important to assure reliable time series of data for comparative studies. Aamlid et al. (2007) recommend conducting vegetation studies at least every 5 years. However, to differentiate between short-term fluctuations and long-term vegetation dynamics, they also recommend conducting studies annually.

Three key components of biodiversity can be recognized in forest ecosystems: composition, structure and function (Schulze and Mooney 1994). The main approach to describing these components is via indicators related to structure and composition that are more feasible to measure and/or estimate (Ferris and Humphrey 1999). Plant life-forms (plant species or species groups), can be described using measures such as frequency, density and structural attributes such as cover and biomass, all of which are area-related measures. All measures can be assessed using NFI plots, points, and transects depending on particular NFI objectives and designs.

3.7.2 Assessing Plant Species and Ground Vegetation in NFIs

Not all countries assess ground vegetation as part of their NFIs; of the 26 European countries contributing data as part of COST Action E43, only 21 assess ground vegetation (Fig. 3.9).

Table 3.13 Ground vegetation monitoring area

	Vegetation		Area (m^2)
Ellenberg and Mueller-Dumbois (1967)	Forest stands (understory)		50–200
	Grasslands		50–200
	Dwarf shrub		10–25
	Bryophytes communities		1–4
	Lichens communities		0.1–1
Roberts-Pichette and Gillespie (1999)	Small tree and shrub layer	Standard	25
		Dense packed	4
	Woody plants (height < 1 m), herbaceous lichens and fungi	Standard	1
		Numerous individuals	0.25
		and dense packed	0.0625
Oosting (1956)	Shrubs height < 3 m		16
	Herb layer		1

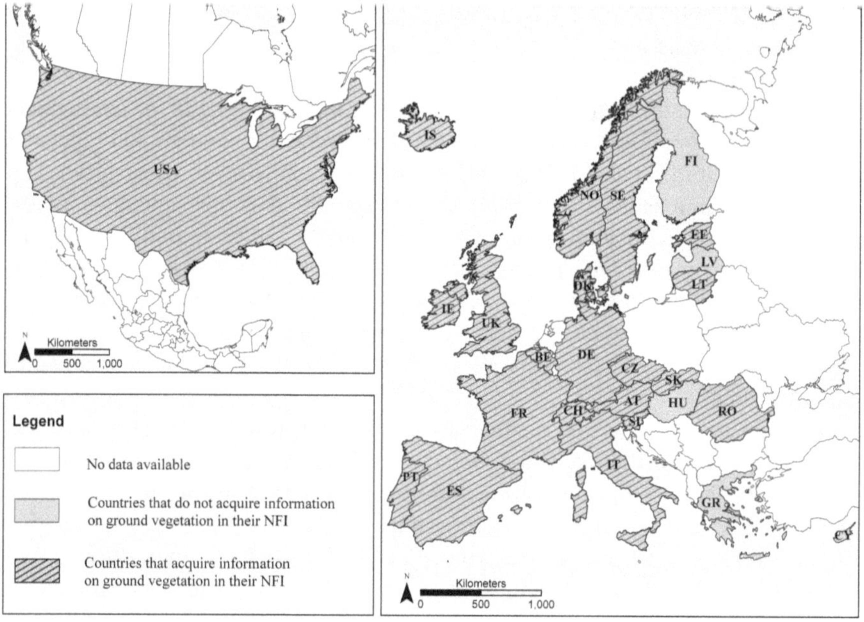

Fig. 3.9 Countries whose NFIs assess ground vegetation

For countries whose NFIs assess ground vegetation, harmonization is extremely difficult because of differences in plot sizes, inventory cycles, sampling seasons, ground vegetation classifications (life-forms), methods of combining different life-forms into categories and the methods used to measure attributes.

3.7.3 Similarities and Differences in the NFIs Ground Vegetation Assessment

As is widely known, the number of species observed depends on the area surveyed. Ground vegetation monitoring areas range from 0.7 to 2,500 m^2, although some of these differences are due to the use of subplots by some NFIs (Table 3.14).

Within an individual NFI, the frequency of ground vegetation inventories is the same for each vegetation type. Six countries have a ground vegetation inventory frequency of 10 years, six have a frequency of 5 years, one has a frequency of 2 years, four countries conduct inventories annually, and in two countries the frequency is not fixed. In general, NFIs do not conduct temporally intensive ground vegetation monitoring programs, although they could assess long-term (10-year) vegetation dynamics.

An additional issue associated with ground vegetation monitoring is whether the assessment season is optimized with respect to assessment requirements. If so, the

Table 3.14 Ground vegetation monitoring area (m^2)

Country	Area (m^2)					
	Shrubs	Herbs	Ferns	Lichens	Liverworts	Mosses
Austria	300.00	300.00	300.00	–	–	300.00
Belgium	452.39	452.39	452.39	–	452.39	452.39
Cyprus	314.16	–	–	–	–	–
Czech Republic	500.00	500.00	500.00	500.00	500.00	500.00
Denmark	314.16	314.16	314.16	314.16	314.16	314.16
Estonia	314.16	–	–	–	–	–
Finland	–	–	–	–	–	–
France	700.00	700.00	700.00	700.00	700.00	700.00
Germany	314.16	314.16	314.16	–	–	314.16
Iceland	0.75	0.75	0.75	0.75	–	0.75
Ireland	500.00	500.00	500.00	500.00	–	500.00
Italy	25.00	–	–	–	–	–
Lithuania	40.00 60.00	–	–	–	–	–
Norway	250.00	–	–	–	–	–
Portugal	Variable. 1/2/4/8/16/32	Variable. 1/2/4/8/16/32	–	–	–	–
Romania	3.14	3.14	–	3.14	–	3.14
Slovakia	499.55	499.55	499.55	499.55	–	499.55
Spain	1,963.50 706.85 314.16	1,963.50 706.85	1,963.50 706.85	1,963.50	1,963.50	1,963.50
Sweden	314.16	0.50 100.00	0.50 100.00	0.50 100.00	0.50 100.00	0.50 100.00
Switzerland	200.00 2,500.00 (total GV cover)	–	–	–	–	–
USA	12 quadrats per plot of 1.00 m^2 each	12 quadrats per plot of 1.00 m^2 each	12 quadrats per plot of 1.00 m^2 each	12 quadrats per plot of 1.00 m^2 each	–	12 quadrats per plot of 1.00 m^2 each

growing season is usually selected as the optimal period, particularly in countries with severe climates. Responses to the second WG3 questionnaire (Section 3.1.1) indicate that nearly half the countries have no fixed period, whereas the other half have a fixed period.

Among the countries whose NFIs collect ground vegetation data, all record information for shrubs, 71% for herb and ferns, 62% for lichens and mosses but only 14% for liverworts (Fig. 3.10).

A very high percentage of countries assess liverworts, lichens, mosses, ferns and herbaceous as a group but do not specify species or genera. For shrubs, the pattern changes completely; most countries uses species lists, five attempt to identify all species present, but five do not consider species (Fig. 3.11). The extent and completeness of these species lists vary so much by country that species richness indices are difficult to use because not all species are measured and because different proportions of the total pool of species are included in the NFIs measurements. In addition, some countries use different levels of genus, and countries do not use the same criteria for recording species even when using the same species list. For example, at least one NFI selects species that cause forest management problems whereas another country only selects dominant species.

Among the countries that assess each ground vegetation life-form (shrubs, ferns, mosses), multiple attributes are measured with coverage being the most common (Fig. 3.12).

Coverage is a very useful and descriptive variable and is used in European networks such as ICP Forests, the International Co-operative Programme on Assessment and Monitoring of Air Pollution Effects on Forests operating under the UNECE Convention on Long-range Transboundary Air Pollution (ICP 2009).

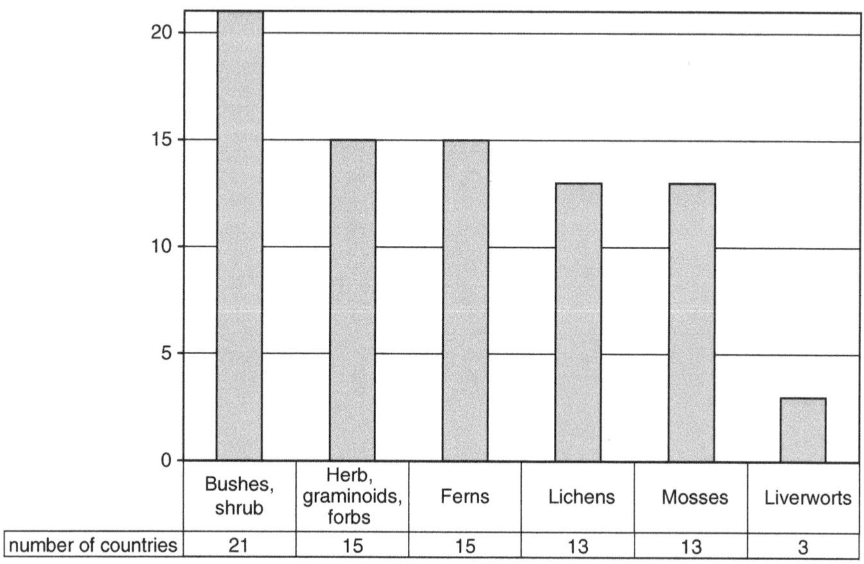

Fig. 3.10 Number of countries assessing ground vegetation life-forms

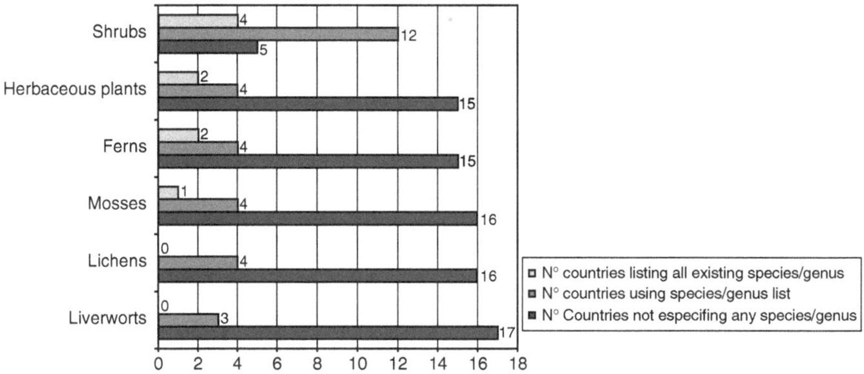

Fig. 3.11 Ground vegetation taxonomical analysis

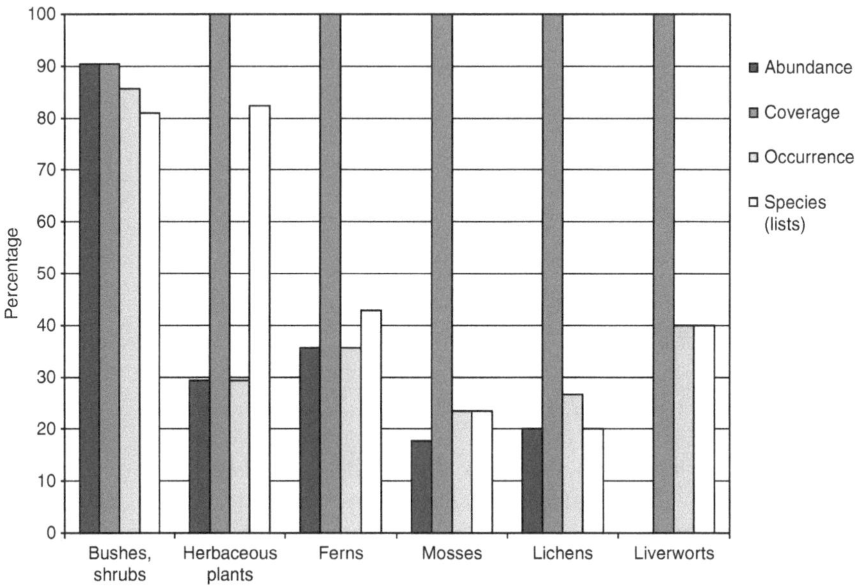

Fig. 3.12 Percentage of countries assessing the different ground vegetation life-forms attributes

Thus, analyses of coverage sampling methods is important (Fig. 3.13). The NFIs of most countries can report coverages by life-form as a group, but only a few can report by species based on species lists or by sub-groups (e.g. graminoids).

For shrub coverage as a group, definitions of layers and cover classifications vary by country. For most NFIs, the lower shrub layer limit ranges from 0.3 to 0.5 m and the upper limit ranges from 5 to 7 m. One-third of NFIs use the Braun-Blanquet scale (Braun-Blanquet 1965); one-third use a modified Braun-Blanquet (1/3) scale; and one-third use percentage scales without classes. Harmonization is particularly difficult when the cover scales are based on wider cover classes.

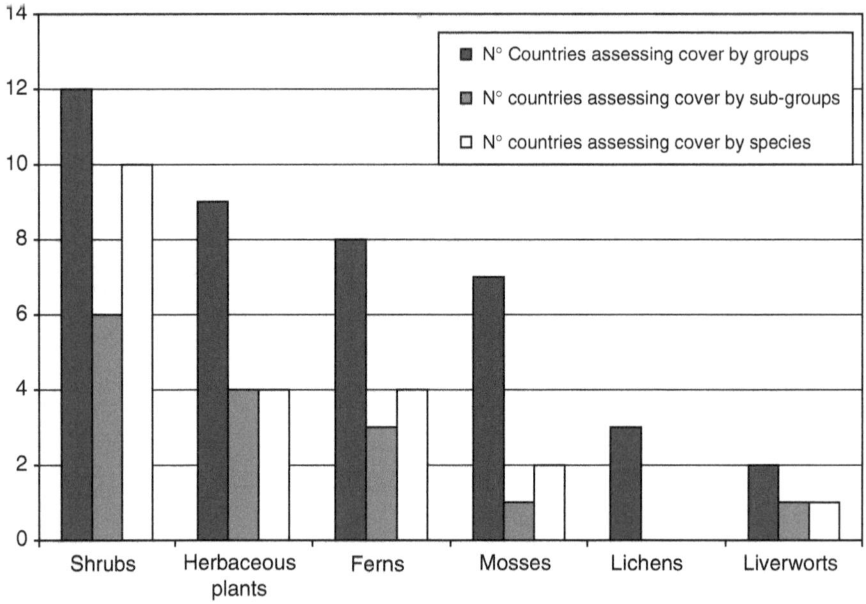

Fig. 3.13 Countries assessing ground vegetation cover and its exhaustiveness

In summary, the shrub group is the mostly commonly sampled (recorded by every country) followed by herbs, ferns mosses, lichens and finally liverworts. Coverage is the most frequently measured attribute for every life-form with the Braun-Blanquet scale or its modifications.

As a result of the large variability in ground vegetation composition, reports in the literature recommend intensive monitoring with at least two visits per year in the growing seasons. Also, different sampling areas are recommended for different ground vegetation life-forms, a feature that decreases NFI costs. Attributes measured and exhaustiveness depend on NFI objectives. Nevertheless, cover is one of the most important variables indicated by the literature, and, fortunately, it can be readily harmonized. In addition, shrubs constitute a ground vegetation group whose attributes can also be easily harmonized because of their perennial character, as opposed to the transient nature of herbaceous plants. Of course, depending on the bio-geographical area, relative abundances of different ground vegetation life-forms or vegetation groups are highly variable. Therefore assessment of every ground vegetation group's cover is important as a means of comparing all the data and studying dynamics.

Developing quantitative information on biodiversity and ecosystem responses to human influence is a high priority. Not all management treatments ensure the maintenance of floristic diversity in the long-term (Gärtner and Reif 2005). Data and the results of ground vegetation analyses can be fed back into the decision-making process using an adaptive management framework (Schulte et al. 2006).

Inclusion of a ground vegetation monitoring component in NFI programs is recommended because it contributes to the completeness of forest biodiversity information. To minimize NFI costs and optimize methods, reliable investigations are needed to determine the minimal number of sampling plots and minimal plot area for each forest type and ground vegetation life-form. The first step is to determine the objectives for ground vegetation monitoring within each NFI because they form the basis for developing species lists. For example, the objective may be to identify the most frequent species, the most important species, or perhaps invasive species. Once the objectives have been formulated, then ground vegetation species lists may be revised and harmonized.

3.7.4 Harmonization Possibilities

Experimental harmonization tests were conducted for shrubs and ground vegetation components using data provided by ten countries (Chap. 4). For the shrub test, harmonization investigations were conducted for two variables, shrub species and cover scales, and two indicators were used, presence/absence of shrub species and shrub species cover. For the ground vegetation components test, harmonization investigations focused on ground vegetation layers, cover scales and sampling season. Results of the harmonization tests are reported in Sect. 5.7.

3.8 Forest Naturalness

3.8.1 Background

Conservation biology has emerged as the application of scientific principles to address the effects of human disturbance on ecological systems (Soulé 1985). This view emphasizes maintenance of the natural integrity of ecosystems and is guided by the principle that naturally evolved ecosystems are superior to disturbed or artificial systems with respect to ecosystem function and biodiversity (Liira et al. 2007). The term *ecological integrity* has been used to ascribe value to older, natural forest stands (Angermeier 2000; Karr 1991; Woodley et al. 1993; Angermeier and Karr 1994; Ohlson et al. 1997). The first reference to integrity in the ecological sense may have been Aldo Leopold's (1949) famous statement, "A thing is right when it tends to preserve the integrity, stability, and beauty of the biotic community." More recently, Frey (1975) stated that integrity is "the capability of supporting and maintaining a balanced, integrated, adaptive community of organisms having a species composition, diversity, and functional organization comparable to that of natural habitat of the region." Angermeier (2000) asserts that the foundation of ecological integrity lies in the concept of naturalness. Definitions of naturalness vary (Šaudytė et al. 2005), but all relate to a continuum with entirely natural and entirely artificial at the extremes (Angermeier 2000) (Table 3.15).

Table 3.15 Categories of forest naturalness from the literature

Category	Description	References
Primeval, pristine, virgin	Forests that have evolved without any human intervention and that have structures corresponding approximately to the climax forest stage	Lindenmayer and Franklin (1997), Angermeier (2000), Šaudytė et al. (2005), Liira et al. (2007)
Near-natural	Naturally regenerated forests composed of native tree species that have been managed in the past but have now experienced a relatively long period of low human interference	Mountford (2002)
Intact	Forests that include all the critical ecosystem components and structure and with processes functioning within normal limits	Anderson (1991)
Semi-natural	Forests with development influenced by human activities using ecological principles with the result that species composition and forest structure are similar to natural forest	Hansen et al. (1991), Šaudytė et al. (2005)
Conventionally managed	Forests that exhibit levels of intense management	Liira et al. (2007)
Plantation	Forests of either native or non-native species, typically with artificial plantings using regular spacing	–

The concept of naturalness has multiples uses: (1) to describe the ecological value of forest ecosystems (Usher 1986; Lane and Tait 1990; Crumpacker 1998; Šaudytė et al. 2005) so that planning and management practices can be applied to maintain those values (Smith and Theberge 1987); (2) to judge management efforts to maintain and conserve biodiversity (Harris 1984; Hansen et al. 1991; Hoerr 1993; Norton 1996; Lähde et al. 1999; Trass et al. 1999; Angermeier 2000; Norden and Appelqvist 2001; Bartha et al. 2006; Liira et al. 2007); and (3) to identify natural, old-growth forests for purposes of establishing protection areas (Smith and Theberge 1987; Gustafsson 2002; Uotila et al. 2002; Branquart and Latham 2007).

Two perspectives on naturalness serve as bases for complementary approaches to its assessment. The first approach is based on an assessment of ecosystem processes (Peterken 1996), and the second approach is based on the degree of human influence (Anderson 1991; Rolston 1990). Jalas (1955) introduced the term *hemeroby*, from the Greek *hemeros* meaning cultivated, tamed, or refined, as a measure of human impact, and Sukopp et al. (1990) and Kowarik (1990) characterized hemeroby as a measure that integrates the effects of past and present human influence on ecosystems.

Although no single variable has been found adequate for assessing all aspects of naturalness, assessment of all variables is not only impossible but would not produce policy-relevant information for non-experts (Branquart and Latham 2007). Thus, for large area analyses such as for Europe or the USA, a small set of appropriate

Table 3.16 Proposed indicators of naturalness from the literature

Categories	Indicators	References
Ecosystem processes	Numbers of old and large trees	McComb et al. (1993), Gustafsson (2002), Nilsson et al. (2002), von Oheimb et al. (2005)
	Number of height or canopy layers	McComb et al. (1993), Uotila et al. (2002)
	Deadwood	McComb et al. (1993), FRA (2000), Gustafsson (2002), Nilsson et al. (2002), Christensen et al. (2005), Liira et al. (2007)
	Shapes of diameter distributions	Koop and Hilgen (1987), Buongiorno et al. (1994), Kuuluvainen et al. (1996), (1998), Linder et al. (1997), Lähde et al. (1999), Maltamo et al. (2000), Uotila et al. (2002), Siipilehto and Siitonen (2004), von Oheimb et al. (2005), Westphal et al. (2006)
	Species composition	Gustafsson and Hallingbäck (1988), McComb et al. (1993), Cochrane and Schulze (1999), FRA (2000), Šaudytė et al. (2005), Bartha et al. (2006)
	Number of microhabitats	Winter and Möller (2008)
	Growing stock volume	Kuuluvainen et al. (1998), Uotila et al. (2002), Liira et al. (2007)
	Forest structural diversity	Hansen et al. (1991), Burschel (1992), McComb et al. (1993), Buongiorno et al. (1994), Larsen (1995), Kuuluvainen et al. (1996), Peterken (1996), Lähde et al. (1999), McComb and Lindenmayer (1999), Uotila et al. (2002), Winter (2005)
Hemeroby	Signs of slash and burn	Uotila et al. (2002)
	Cut stumps	FRA (2000), Uotila (2002), Šaudytė et al. (2005)
	Deadwood	Uotila et al. (2002)
	Growing stock volume	Uotila et al. (2002)
	Signs of silvicultural management	–

indicators is required, although a rather large number has been proposed in the literature (Table 3.16).

3.8.2 Assessing Forest Naturalness in NFIs

Although naturalness is a more familiar concept in Europe than elsewhere, even in Europe, little natural forest remains. Branquart and Latham (2007) report that ratios of the areas of nearly natural forest to the total areas of forest and other wooded lands for Europe are 0.001 for western Europe, 0.013 for southern Europe, 0.025 for central Europe, and 0.083 for northern Europe. They attribute the greater ratio for northern Europe to the chronologically later onset of human influence. Not all countries assess naturalness as part of their NFIs; of the 19 European countries

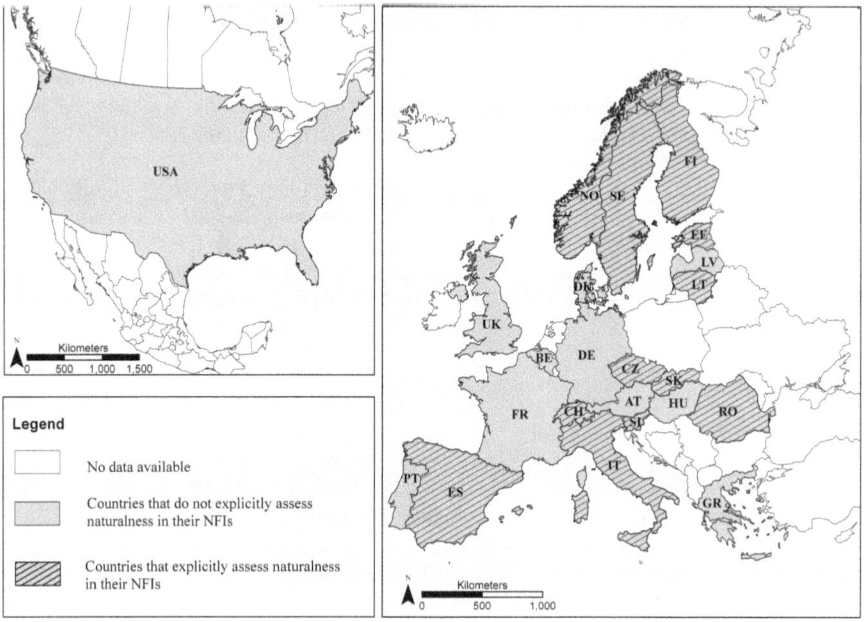

Fig. 3.14 Countries whose NFIs assess forest naturalness

contributing data as part of COST Action E43, only 13 explicitly assess naturalness (Fig. 3.14). Generally, countries with the least amount of forest area characterized as natural or close-to-natural, are the countries that do not explicitly assess naturalness, although Italy and Spain may be exceptions.

3.8.3 Similarities and Differences in the NFIs Forest Naturalness Assessment

For countries whose NFIs assess forest naturalness, harmonization is difficult because of differences in assessment criteria and differences in categories of naturalness. Among the 11 countries that assess naturalness, five used specific observations and five used traditional NFI data. The most commonly used criterion is signs of silvicultural management although species composition, tree or stand age, and presence and/or amount of deadwood are also commonly used (Fig. 3.15).

Approximately 80% of countries use one or two criteria, whereas approximately 20% use more than two criteria with Estonia and Sweden each using five. The latter two countries both use tree or stand age, deadwood and signs of silviculture. In addition, Estonia records the water regime disturbance and Sweden assesses stand structure with respect to diameter distribution and number of vertical layers. In addition, to using different assessment criteria, NFIs also use different categories of forest naturalness (Table 3.17).

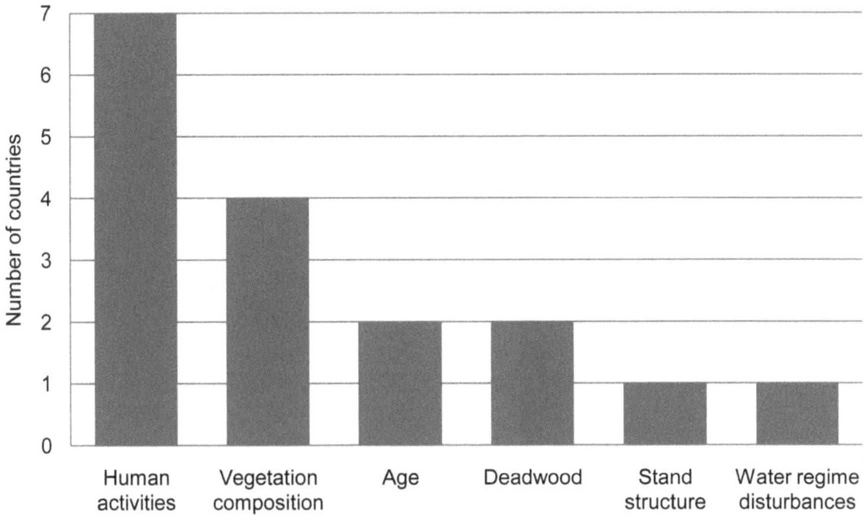

Fig. 3.15 Primary criteria used for assessing forest naturalness (Czech Republic, Estonia, Finland, Italy, Lithuania, Norway, Romania, Spain, Sweden, Switzerland)

Reports in the literature of attempts to construct objective, quantitative indices of forest naturalness are few and specifically adapted to individual cases (Machado 2004). Although NFIs base their assessments on observations of a variety of variables, the assessments are often visual or subjective. Scholes and Biggs (2005) proposed a biodiversity intactness index, but it is still subjective because it relies heavily on expert judgments. Petriccione (2006) proposed a more quantitative approach based on comparing observed and potential vegetation types. If the two types are different, a naturalness value of 0 is assigned; if they are the same, then the naturalness value is based on the average of six indices characterizing disturbance and species types, richness, and coverage. For beech forests in Germany, (Winter et al. 2005; Winter 2006) reported three indices and average thresholds for distinguishing managed and near-natural forest: (1) 1.8 living trees per ha with dbh≥80 cm, albeit with diminished vitality; (2) more than half the standing and fallen deadwood pieces with length≥10 m; and (3) three trees per ha with dbh≥60 cm and with the fungus *Fomes fomentarious.*

Despite the difficulty in assessing naturalness, international interest in silvicultural practices that mimic natural forest ecosystem processes has increased in recent years (Lähde et al. 1999). In most Central and Mediterranean European countries, a close-to-nature silvicultural approach has been postulated as the most appropriate management strategy (Ciancio et al. 1999; von Oheimb et al. 2005), and forest ecosystems with a high degree of naturalness are becoming the forest management standard for comparison (Peterken 1996; Lähde et al. 1999). In addition, European countries have committed to implementing guidelines for sustainable forest management that conserve biodiversity and promote naturalness (Loiskekoski et al. 1993; Ratcliffe and Peterken 1995). Therefore, there exists ample rationale for

Table 3.17 European forest naturalness categories

Countries	Categories	Descriptions
Czech Republic	–	Naturalness is assessed by comparing current tree species composition on the plot and the potential natural tree vegetation on the plot
Estonia	Natural	Five conditions: (1) age (mixed-aged stand); if even-aged stand, then conifers forest age must be at least 100 years, deciduous forest 80 years (2) mixed species stand (3) deadwood (5–10% of growing trees number) (4) water regime is undisturbed by human (5) cutting signs are not established (or 100 years passed since logging)
Finland	Undisturbed	No sign of human disturbance
	Minor disturbance I	Some signs of human impact other than forestry
	Minor disturbance II	Some signs of forest treatments a long time ago
	Clear disturbance I	Impact of forestry has decreased naturalness
	Clear disturbance II	Impact of other human activity has decreased naturalness (e.g. agriculture)
Italy [a]	Undisturbed	No human disturbance at all or for a long time
	Semi-natural	If forests are or were disturbed
	Artificial	Artificial forest (plantations included)
Lithuania	Natural stands	Planted trees comprise less than 20% of stand volume
	Mixed stands	Planted trees comprise 20–50% of stand volume
	Planted stands	Planted trees comprise more than 50% of stand volume
	Plantings	Planted stands up to 10 years
Norway	Undisturbed	Natural forest dynamics; large enough to maintain its natural characteristics; no known significant human intervention or where the last significant human intervention was long enough ago to have allowed the natural species composition and processes to have become re-established (Michalak 2008).
	Semi-natural	Neither undisturbed by man nor plantation (Michalak 2008).
	Plantations	Established by planting or/and seeding in the process of afforestation or reforestation; planted/seeded trees are intensively managed stands that are one/two species at plantation or even age class or of regular spacing (Michalak 2008).
Romania	Natural fundamental	With the same or almost identical composition to the natural potential vegetation and with natural regeneration.
	Derived stands	Two subclasses: partially altered and totally altered
	Artificial stands	Artificial regeneration and different types of composition
	Indefinite young stands	Forest that cannot be included into the first three categories due to small age and insufficient information on the future forest stand evolution
Slovakia	Virgin forest	No human impact, native species, auto-regulation functional processes, dead wood (several decay stages)

(continued)

Table 3.17 (continued)

Countries	Categories	Descriptions
	Natural forest	More than 85% of the parameters of the virgin forest, natural tree species composition, spatial structure modified but auto-regulation functional processes, close to nature management
	Semi-natural forest	50–85% of the parameters of the virgin forest, no invasive species
	Changed forest	Clear human impact, native or non-native tree species, clearcut system, generally one layer and regular spacing
	Artificial forest	Artificial origin of trees, introduced species, generally one species, regular spacing
Slovenia	No anthropogenic influences	Forests with natural tree structure, natural regeneration…, where is no sign of forest management (virgin forest, forest reserve, protective forest…)
	Close to nature manage forest	–
	Forests with exchanged tree structure	Where more than 90% of growing stock belongs to non-native tree species
Spain	Natural	Forest stand with no anthropogenic origin coming from natural regenerating (seed scattered by the stand, stock shoot or root shoot)
	Artificial	Forest stand with anthropogenic origin coming from regeneration by planting (plant from nurseries) or by seeding (scattered by man)
	Naturalized	Artificial forest stand (with anthropogenic origin) which has lost the regularity, the stand structure has became irregular and natural regeneration of different species exists (period 30–50 years depending on the area)
Sweden[b]	Old-growth	>150 years, cwd, no forestry measures the last 25 years, uneven-aged, large diameter variation, at least two layers
	Plantation forest	Even-aged, small diameter variation, one layer, monoculture, strict geometric distribution
	Normal forest	Other
Switzerland[c]	Near to nature	Broadleaved forest areas only: stands with less than 10% or 25% of coniferous trees (depending on the plant community)
	Fairly far from nature	Broadleaved forest areas only: forests with up to 75% coniferous trees
	Far from nature	Broadleaved forest areas only: forests with over 75% coniferous trees
	Very far from nature	Broadleaved forest areas only: stands where the proportion of spruce alone is over 75%

[a] In addition, origin of ground vegetation (regeneration and shrubs): natural, artificial, coppice

[b] In addition, continuous tree coverage without species change since eighteenth century, >100 years, stocking above 30%: no, possibly, yes

[c] No judgement on the closeness to nature could be made in the natural area of coniferous forests, though as a whole, they must be considered as fairly near to nature since very often the site conditions dictate the composition of species (e.g. in the Alps at high altitudes)

investigating possibilities for an objective approach to assessing natural forest that can be used to harmonize estimates for countries whose NFI features, definitions, measurement protocols, assessment criteria, and classes of natural forest differ.

As a conclusion, the NFIs of approximately 60% of responding countries conduct naturalness assessments, mainly based on a specific classification; the NFIs of only three countries compare current and potential natural vegetation. The primary differences are twofold: definitions differ, even if the same terms are used, and the number of naturalness classes varies, generally between three and five. The similarities pertain to the variables used to define naturalness classes, primarily, age structure, regeneration types, tree origin and cuttings.

3.8.4 Harmonization Possibilities

Harmonization of assessments of forest naturalness is difficult for multiple reasons including the lack of a common definition, lack of well-accepted measures and indices, and lack of natural forests in much of Europe as a motivating factor. The analyses using the common database (Chap. 4) have three objectives: (1) to assess the utility of NFI data for formulating indicators of forest naturalness, (2) to develop an objective method for assessing forest naturalness, and (3) to investigate the effects of different NFI features such as plot configurations and measurement methods on estimates of the indicators and assessment methods. The results of the analyses are reported in Sect. 5.8.

3.9 Summary

After selection of the 13 biodiversity variables (Sect. 2.3) based on their importance and feasibility for assessment by NFIs, responses were solicited from participating countries regarding the degree to which the variables are now assessed. Two conclusions were evident: (1) most countries currently assess most of the variables, but (2) consensus is lacking on assessment methods and necessary field crew expertise, suggesting that harmonization would require emphasis on field operations.

For each of the seven essential features into which the 13 variables were grouped, more detailed assessments were conducted. For forest categories, the conclusion was that the only major difference in classification systems used by European NFIs was whether potential or actual vegetation was used to define classes. Thus, the prospects for harmonization of forest categories are considered excellent. For forest structure, the prospects depend on the variable. For tree species, the prospects are excellent because the variable is assessed in the same manner by all NFIs. For dbh and height, considerable variability in measurement thresholds were found, but otherwise the harmonization prospects are good.

For social position, definitions of classes varied, but harmonized estimates of proportions for dominant, intermediate, and suppressed classes are considered possible. Prospects for harmonized estimates of layers are considerably poorer because of different definitions, thresholds and the uncertainty associated with visual assessment methods.

Harmonized estimation of forest age is impeded by the increasing proportion of uneven-aged stands for which age is often not assessed, different definitions, and different assessment methods. However, agreement on dominant age as a reference definition would greatly increase the prospects. Deadwood is becoming an increasingly popular indicator of sustainable forest management. Unfortunately, considerable variability was found in deadwood definitions, components (e.g., stumps, limbs), sampling methods, and measurement thresholds. Thus, harmonized deadwood estimation will require development of bridges. Harmonization of regeneration estimates faces challenges due to differences in assessment approaches such as presence/absence versus coverage and all species versus dominant species. Harmonized estimation may be restricted to change in regeneration success. Harmonized estimation for ground vegetation also faces serious challenges due to differences in the components assessed (e.g., small trees, shrubs, herbs, bryophytes, lichens), difference in height thresholds, and differences in categories for which ground vegetation is reported.

Forest naturalness integrates many of the other essential features. However, many countries do not assess naturalness, and among those that do, assessment variables, methods, and reporting classes vary considerably. For harmonized assessment using NFI variables, the hemeroby approach, which emphasizes indications of human influence, is extremely sensitive to plot size. Harmonization using the ecosystem processes approach requires a common dbh threshold and similar plot sizes.

The overall conclusion is that harmonization will be considerably easier for some essential features than for others. The factors leading to difficulties often are related to different definitions, different reporting classes, different measurement thresholds, and different features of sampling protocols such as plot sizes and configurations. Nevertheless, construction of reference definitions and bridges greatly facilitate harmonization for all essential features as is illustrated in Chap. 5.

References

Aamlid, D., Canullo, R., & Starlinger, F. (2007). Manual on methods and criteria for harmonized sampling, assessment, monitoring and analysis of the effects of the effects of air pollution on forest. Part VIII. Assessment of ground vegetation. International co-operative programme on assessment and monitoring of air pollution effects on forests, United Nations Economic Commission for Europe (UNECE) Convention on long-range transboundary air pollution. http://www.icp-forests.org/pdf/manual8.pdf. Accessed Mar 2009.

Aguilo, M., Aramburu, M. P., Blanco, A., Calatayud, T., Carrasco, R., Castilla, G., Castillo, V., Cenal, M., Cifuentes, P., Diaz, M., Diaz, A., Escribano, R., Escribano, M., Frutos, M., Galiana, F., García, A., Glaria, G., González, S., González, C., Iglesias, E., Martin, A., Martinez, E.,

Milara, R., Monzon, A., Ortega, C., Otero, I., Pedraza, J., Pinedo, A., Puig, J., Ramos, A., Rodríguez, I., Sanz, M. A., Tevar, G., Torrecilla, I., & Yoldi, L. (1992). *Guía para la elaboración de estudios del medio físico: contenido y metodología* (p. 809). Madrid: Ministerio de Medio Ambiente.

Alberdi, I., Condes, S., & Martinez-Millán, J. (2010). Review of monitoring and assessing ground vegetation biodiversity in national forest inventories. *Environmental Monitoring and Assessment, 164*, 649–676.

Aldrich, P. R., Parker, G. R., Romero-Severson, J., & Michler, C. H. (2005). Confirmation of oak recruitment failure in Indiana old-growth forest: 75 years of data. *Forest Science, 51*, 406–416.

Anderson, J. E. (1991). A conceptual framework for evaluating and quantifying naturalness. *Conservation Biology, 5*(3), 347–352.

Angermeier, P. L. (2000). The natural imperative for biological conservation. *Conservation Biology, 14*(2), 373–381.

Angermeier, P. L., & Karr, J. R. (1994). Biological integrity versus biological diversity as policy directives. *Bioscience, 44*(10), 690–697.

Augusto, L., Dupouey, J.-L., & Ranger, R. (2003). Effects of tree species on understory vegetation and environmental conditions in temperate forests. *Annals of Forest Science, 60*, 823–831.

Avery, T. E., & Burkhardt, H. E. (2002). *Forest measurements* (5th ed., p. 456). Boston: McGraw-Hill.

Axelsson, A. L., & Fridman, J. (2005). Country report for forest focus- Common project. http://www.resgeom.slu.se/resana/projekt/common/Documents/Country%20reports/Sweden%20country%20report.pdf. Accessed Mar 2009.

Axelsson, R., Angelstam, P., & Svensson, J. (2007). Natural forest and cultural woodland with continuous tree cover in Sweden: how much remains and how is it managed? *Scandinavian Journal of Forest Research, 22*, 545–558.

Baier, R., Meyer, J., & Göttlein, A. (2005). Regeneration niches of Norway spruce (Picea abies [L.] Karst.) saplings in small canopy gaps in mixed mountain forests of the Bavarian Limestone Alps. *European Journal of Forest Research, 126*, 11–22. Published online: 6 Dec 2005 Springer.

Barbati, A., Corona, P., & Marchetti, M. (2007). A forest typology for monitoring sustainable forest management: the case of European forest types. *Plant Biosystems, 141*, 93–103.

Barsoum, N. (2002). Relative contributions of sexual and asexual regeneration strategies in *Populus nigra* and *Salix alba* during the first years of establishment on a braided gravel bed river. *Evolutionary Ecology, 15*, 255–279.

Barsoum, N., Muller, E., & Skot, L. (2004). Variations in levels of clonality among *Populus nigra* L. stands of different ages. *Evolutionary Ecology, 18*, 601–624.

Bartha, D., Odor, P., Horvath, T., Timar, G., Kenderes, K., Standovar, T., Bölöni, J., SZmorad, F., Bodonczi, L., & Aszalos, R. (2006). Relationship of tree stand heterogeneity and forest naturalness. *Acta Silvatica et Lignaria Hungarica, 2*, 7–22.

Bitterlich, W. (1984). *The relascope idea. Relative measurements in forestry*. Norwich: Commonwealth Agricultural Bureaux, Page Bros.

Blume, P., & Sukopp, H. (1976). Ökologische Bedeutung anthropogener Bodenveränderungen. *Schreiftenreihe Vegetationskunde, 10*, 7–89.

Boddy, L. (2001). Fungal community ecology and wood decomposition processes in angiosperms: from standing tree to complete decay of coarse woody debris. *Ecological Bulletin, 49*, 43–56.

Böhl, J., & Brändli, U.-B. (2007). Deadwood volume assessment in the third Swiss National Forest Inventory: methods and first results. *European Journal of Forest Research, 126*, 449–457.

Bohn, U., Gollub, G., & Hettwer, C. (2000). *Map of the natural vegetation of Europe*. Bonn: Federal Agency for Nature Conservation.

Bonham, C. D. (1989). *Measurements for terrestrial vegetation*. United States of America: Wiley.

Bosch, P., & Söderbäck, E. (1997). European environmental state indicators. Project report, European Environment Agency, Copenhagen and Swedish Environmental Protection Agency, Stockholm.

Brändli, U.-B., Bühler, C., & Zangger, A. (2007a). Waldindikatoren zur Artenvielfalt - Erkenntnisse aus LFI und BDM Schweiz. *Schweizerische Zeitschrift fur Forstwesen, 158*, 243–254.
Brändli, U.-B., Bühler, C., & Zangger, A. (2007b). Biodiversität und Waldinventuren. *LFI info, 7*, 1–6.
Branquart, E., & Latham, J. (2007). Selection criteria for protected forest areas dedicated to biodiversity conservation in Europe. In G. Frank et al. (Eds.), *COST Action E27, Protected forest areas in Europe – analysis and harmonization (PROFOR): Results, conclusions, and recommendations*. Vienna, Austria: Federal Research and Training Centre for Forests, Natural Hazards and Landscape (BFW).
Braun-Blanquet, J. (1965). *Plant sociology; the study of plant communities*. London: Halner.
Brodie, A., Bowering, M., Jaross, W., Reimer, D., & Lu, B. (2007). Combining management goals of wildlife habitat conservation and revenue on Washington State trusts U S forest service Pacific Northwest research station general technical report PNW-GTR Issue: 67–80.
Buongiorno, J., Dahir, S., Ly, H.-C., & Lin, C.-R. (1994). Tree size diversity and economic returns in uneven-aged forest stands. *Forest Science, 40*(1), 83–103.
Burschel, P. (1992). Experiments in mixed mountain forests in Bavaria. In M. J. Kelty, B. C. Larson, & C. D. Oliver (Eds.), *The ecology and silviculture of mixed-species forests* (pp. 183–215). Dordrecht: Kluwer Academic Publishers.
Chirici, G., Winter, S., Bastrup-Birk, A., Rondeux, J., Bertini, R., McRoberts R.E., Barsoum, N., Alberdi, I., Brändli, U.-B., & Marchetti, M. (submitted). Harmonised estimation of forest biodiversity indicators at the cross-regional scale using data from national forest inventories. Forest Science.
Christensen, M., Hahn, K., Mountford, E., Ódor, P., Rozenberger, D., Diaci, J., Standovar, T., Wijdeven, S., Winter, S., Vrska, T., & Meyer, P. (2005). Dead wood in European beech (Fagus) forest reserves. *Forest Ecology and Management, 210*, 267–282.
Ciancio, O., Corona, P., Iovino, F., Menguzzato, G., & Scotti, R. (1999). Forest management on a natural basis: the fundamentals and case studies. *Journal of Sustainable Forestry, 1*(2), 89–95.
Cochrane, M. A., & Schulze, M. D. (1999). Fire as a recurrent event in tropical forests of the eastern Amazon: Effects on forest structure, biomass and species composition. *Biotropica, 31*, 2–16.
COST E43 (2005). Harmonization of National forest inventories in Europe: Techniques for common reporting. Reports and proceedings. http://www.metla.fi/eu/cost/e43. Accessed Mar 2008.
Crumpacker, D. W. (1998). Prospects for sustainability of biodiversity based on conservation biology and US Forest Service approaches to ecosystem management. *Landscape and Urban Planning, 40*, 47–71.
Davies, C. E., Moss, D., O. & Hill M. (2004). EUNIS habitat classification revised 2004. http://EUNIS.eea.eu.int/upload/EUNIS_2004_report.pdf. Accessed Mar 2009.
Deconchat, M., & Balent, G. (2001). Vegetation and bird community dynamics in fragmented coppice forests. *Forestry, 74*, 105–118.
De Graaf, R. M., Hestbeck, J. B., & Yamasaki, M. (1998). Associations between breeding bird abundance and stand structure in the White Mountains, New Hampshire and Maine, USA. *Forest Ecology and Management, 103*, 217–231.
Douhovnikoff, V., Cheng, A. M., & Dodd, R. S. (2004). Incidence, size and spatial structure of clones in second-growth stands of coast redwood, *Sequoia semperverins* (Cupressaceae). *American Journal of Botany, 91*, 1140–1146.
Duelli, P., & Obrist, M. K. (2003). Biodiversity indicators: the choice of values and measures. *Agriculture, Ecosystems & Environment, 98*, 87–89.
Dusan, R., Mikac, S., Anić, I., & Diaci, J. (2007). Gap regeneration patterns in relationship to light heterogeneity in two old-growth beech – fir forest reserves in South East Europe. *Forestry, 80* (4), 431–443.
Ellenberg, H., & Mueller-Dumbois, D. (1967). A key to Raunkiaer plant life forms with revised subdivisions. Berichte des Geobotanisches Institut, Die Eidgenössische Technische Hochschule Stiftung Rübel, Zurich:56–73.
Ellison, A. M., Bank, M. S., Clinton, B. D., Colburn, E. A., Elliott, K., Ford, C. R., Foster, D. R., Kloeppel, B. D., Knoepp, J. D., Lovett, G. M., Mohan, J., Orwig, D. A., Rodenhouse, N. L.,

Sobczak, W. V., Stinson, K. A., Stone, J. K., Swan, C. M., Thompson, J., Von Holle, B., & Webster, J. R. (2005). Loss of foundation species: consequences for the structure and dynamics of forested ecosystems. *Frontiers in Ecology and the Environment, 3*, 479–486.
European Environment Agency (EEA) (2003). An inventory of biodiversity indicators in Europe, 2002. European Environment Agency, Copenhagen. Technical report no 92. pp. 42.
European Environment Agency (EEA) (2006). European forest types. Categories and types for sustainable forest management and reporting. EEA Technical report no. 9/2006.
Eyre, T. J., Kelly, A. L., & Nelder, V. J. (2006). Methodology for the establishment and survey of reference sites for BioCondition. http://www.epa.qld.gov.au/publications/p01993aa.pdf/Methodology_for_the_establishment_and_survey_of_reference_sites_for_BioCondition_version_14.pdf. Accessed Mar 2009.
Ferris, R., & Humphrey, J. W. (1999). A review of potential biodiversity indicators for application in British forests. *Forestry, 72*(4), 312–328.
Flade, M., Möller, G., Schumacher, H., & Winter, S. (2004). Naturschutzstandards für die Bewirtschaftung von Buchenwäldern im nordostdeutschen Tiefland. *Der Dauerwald – Zeitschrift für naturgemäße Waldwirtschaft, 29*, 15–28.
Foley, J. A., DeFries, R., Asner, G. P., Barford, C., Bonan, G., & Carpenter, S. R. (2005). Global consequences of land use. *Science, 309*, 570–574.
Food and Agriculture Organization of the United Nations (FAO) (2005). *Global forest resources assessment update*. Terms and definitions (Final version). Working paper 83. Rome 2004. Available at http://www.fao.org/docrep/007/ae156e/AE156E00HTM. Accessed December 2010.
Frey, D. (1975). Biological integrity of water: An historical perspective. In R. K. Ballentine & L. J. Guarraia (Eds.), *The integrity of water* (pp. 127–139). Washington, DC: EPA.
Gärtner, S., & Reif, A. (2005). The response of ground vegetation to structural change during forest conversion in the southern Black Forest. *European Journal of Forest Research, 124*, 221–231.
Gilg, O. (2005). Old-growth forests: Characteristics, conservation and monitoring. Technical report no 74 bis. ATEN, Montpellier. 96 pp.
Grabherr, G., Koch, G., Kirchmeir, H., & Reiter, K. (1998). *Hemerobie österreichischer Waldökosysteme* (17th ed., p. 493). Innsbruck: Veröffentlichungen des Österreichischen MaB-Programms.
Granke, O. (2006). Assessment of ground vegetation. ForestBIOTA work report. http://www.forestbiota.org. Accessed Mar 2009.
Groombridge, B., & Jenkins, M. D. (1996). *Assessing biodiversity status and sustainability* (Vol. 5). Cambridge: World Conservation Press, WCMC Biodiversity.
Gustafsson, L. (2002). Presence and abundance of red-listed plant species in Swedish forests. *Conservation Biology, 16*(2), 377–388.
Gustafsson, L., & Hallingbäck, T. (1988). Bryophyte flora and vegetation of managed and virgin coniferous forests in South-West Sweden. *Biological Conservation, 44*, 238–300.
Halpern, C. B., & Spies, T. A. (1995). Plant species diversity in natural and managed forests of the Pacific Northwest. *Ecological Applications, 5*, 913–934.
Hansen, A. J., Spies, T. A., Swanson, F. J., & Ohmann, J. L. (1991). Conserving biodiversity in managed forests. *Bioscience, 41*(6), 382–392.
Hanski, I., & Hammond, P. (1995). Biodiversity in boreal forests. *Trends in Ecology & Evolution, 10*, 5–6.
Harris, L. D. (1984). *The fragmented forest*. Chicago: University of Chicago Press.
Helms, J. A. (Ed.). (1998). *The dictionary of forestry*. Bethesda: Society of American Foresters, ISBN 0-939970-73-2. pp. 210.
Hočevar, M., & Kovač, M. (2004). A short description of the Slovenian NFI, updated 27.10.2004. http://www.gozdis.si/departments/geomatics/nfi_si.pdf. Accessed March 2009.
Hoerr, W. (1993). The concept of naturalness in environmental discourse. *Natural Areas Journal, 13*(1), 29–32.
Hummel, F. (1991). Comparisons of forestry in Britain and mainland Europe. *Forestry, 64*(2), 141–155.

International Co-operative Programme on Assessment and Monitoring of Air Pollution Effects on Forests in co-operation with the European Commission (ICP) (2009). http://www.icp-forests.org. Accessed Mar 2010.

Jalas, J. (1955). Hemerobe und hemerochore Pflanzenarten. Ein terminologischer Reformversuch. *Acta Societatia Pro Fauna et Flora Fennica, 72*, 1–15.

Johnson, S., Erika, E., Mudrak, L., & Waller, D. M. (2006). A comparison of sampling methodologies for long-term forest vegetation monitoring in the great lakes network National Parks. http://science.nature.nps.gov/im/units/GLKN/Veg%20Plot%20Comparison.pdf. Great Lakes Network Report. Accessed Mar 2009.

Jonsson, B. G., & Jonsell, M. (1999). Exploring potential biodiversity indicators in boreal forests. *Biodiversity and Conservation, 8*, 1417–1433.

Karr, J. R. (1991). Biological integrity: a long neglected aspect of water resource management. *Ecological Applications, 1*, 66–84.

Khanina, L., Bobrovsky, M., Komarov, A., & Mikhajlov, A. (2007). Modeling dynamics of forest ground vegetation diversity under different forest management regimes. *Forest Ecology and Management, 248*(1/2), 80–94.

Koop, H., & Hilgen, P. (1987). Forest dynamics and regeneration mosaic shifts in unexploited beech (Fagus sylvatica) stands at Fontainebleau (France). *Forest Ecology and Management, 20*, 135–150.

Kowarik, I. (1990). Natürlichkeit, Naturnähe und Hemerobie als Bewertungskriterien. In H. Sukopp, S. Hejný, & I. Kowarik (Eds.), *Urban ecology*. The Hague: SBP Academic Publications.

Kraft, G. (1884). *Beiträge zur Lehre von den Durchforstungen, Schlagstellungen und Lichtungshieben*. Hannover: Klindworth Verlag.

Kuuluvainen, T., Penttinen, A., Leinonen, K., & Nygren, M. (1996). Statistical opportunities for comparing stand structural heterogeneity in managed and primeval forests: an example from boreal spruce forest in southern Finland. *Silva Fennica, 30*(2/3), 315–328.

Kuuluvainen, T., Kimmo, S., & Kalliola, R. (1998). Structure of a pristine Picea abies forest in northeastern Europe. *Journal of Vegetation Science, 9*, 563–574.

Lähde, E., Laiho, O., & Norokorpi, Y. (1999). Diversity-oriented silviculture in the boreal zone of Europe. *Forest Ecology and Management, 118*, 223–243.

Lane, A., & Tait, J. (1990). *Practical conservation woodlands* (p. 128). London: Hodder and Stoughton.

Larsen, J. B. (1995). Ecological stability of forests and sustainable silviculture. *Forest Ecology and Management, 73*, 85–96.

Larsson, T-B, Svensson, L., Angelstam, P., Balent, G., Barbati, A., & Bijlsma, R.-J., Boncina, A., Bradshaw, R., Bücking, W., Ciancio, O., Corona, P., Diaci, J., Dias, S., Ellenberg, H., Fernandes, F. M., Fernàndez-Gonzalez, F., Ferris, R., Frank, G., Møller, P. F., Giller, P. S., Gustafsson, L., Halbritter, K., Hal, S., Hansson, L., Innes, J., Jactel, H., Keannel Doppertin, M., Klein, M., Marchetti, M., Mohren, F., Niemelä, P., O'Halloran, J., Rametsteiner, E., Rego, F., Scheidegger, C., Scotti, R., Sjöberg, K., Spanos, I., Spanos, K., Standovár, T., Tømmerås, Å., Trakolis, D., Uuttera, J., Walsh, P. M., Vandekerkhove, K., Watt, A. D., VenDenMeersschaut, D. (2001). *Biodiversity evaluation tools for European forests*. Ecological Bulletins, 50. 236 pp.

Leibundgut, H. (1956). Empfehlungen für die Baumklassenbildung und Methodik bei Versuchen über die Wirkung von Waldpflegemassahmen. IUFRO Section 23, Report 10.

Leopold, A. (1949). *A sand county almanac*. New York: Oxford University Press.

Liira, J., Sepp, T., & Parrest, O. (2007). The forest structure and ecosystem quality in conditions of anthropogenic disturbance along productivity gradient. *Forest Ecology and Management, 250*, 34–36.

Lindenmayer, D. B., & Franklin, J. F. (1997). Managing stand structure as part of ecologically sustainable forest management in Australian mountain ash forests. *Conservation Biology, 11*(5), 1053–1068.

Linder, P., Elfving, B., & Zackrisson, O. (1997). Stand structure and successional trends in virgin boreal forest reserves in Sweden. *Forest Ecology and Management, 98*, 17–33.

Loiskekoski, M., Mahonen, M., Puranen, R., & Rizk, N. (1993). *Sound forestery – sustainabel development*. (p. 186). Ministry of Agriculture and Foresty. Helsinki, Finland.
Machado, A. (2004). An index of naturalness. *Journal for Nature Conservation, 12*, 95–110.
MacMahon, J. A., Schimpf, D. H., Andersen, D. C., Smith, K. G., & Bayn, R. L. (1981). An organism-centered approach to some community and ecosystem concepts. *Journal of Theoretical Biology, 88*(2), 287–307.
Maltamo, M., Kangas, A., Uuttera, J., Tornianianen, T., & Saramäki, J. (2000). *Forest Ecology and Management, 133*, 263–274.
McComb, W., & Lindenmayer, D. (1999). Dying, dead and down trees. In M. L. Hunter (Ed.), *Maintaining biodiversity in forest ecosystems*. Cambridge: Cambridge University Press.
McComb, W. C., Spies, T. A., & Emmingham, W. H. (1993). Douglas-fir forests: managing for timber and mature-forest habitat. *Journal of Forestry, 91*(12), 31–42.
McGee, G. G., & Kimmerer, R. W. (2002). Forest age and management effects on epiphytic bryophyte communities in Adirondack northern hardwood forests, New York, USA. *Canadian Journal of Forest Research, 32*, 1562–1576.
McRoberts, R. E. (2009). A two-step nearest neighbors algorithm using satellite imagery for predicting forest structure within species composition classes. *Remote Sensing of Environment, 113*, 532–545.
McRoberts, R. E., Tomppo, E. O., Schadauer, K., Vidal, C., Ståhl, G., Chirici, G., Lanz, A., Cienciala, E., Winter, S., & Smith, W. B. (2009). Harmonizing national forest inventories. *Journal of Forestry, 107*, 179–187.
McRoberts, R. E., Winter, S., & Chirici, G. (submitted). Assessing forest naturalness. Forest Science.
Michalak, R. (2008). Comparison of the scope, terms, definitions and classifications applied for the FAO global forest resources assessment 2010 and the MCPFE/UNECE/FAO Report on State of Europe's Forests 2007. Part I definitions and classifications structured according to FRA reporting tables. UNECE/FAO Timber Section. Geneva, August 2008. http://timber.unece.org/fileadmin/DAM/meetings/5.1_Formats_and_definitions_1_01.pdf. Accessed Mar 2009.
Michel, A., & Winter, S. (2009). Tree microhabitat structures as indicators of biodiversity in Douglas-fir forests of different stand ages and management histories in the Pacific Northwest, USA. *Forest Ecology and Management, 257*, 1453–1464.
Ministerial Conference on the Protection of Forests in Europe (MCPFE) (1998). Third ministerial conference on the protection of forests in Europe. 2–4 June 1998, Lisbon/Portugal. Annex 1 of the Resolution L2. Pan-European Criteria and Indicators for Sustainable Forest Management. http://www.mcpfe.org. Accessed Mar 2009.
Ministerial Conference on the Protection of Forests in Europe (MCPFE) (2003). Improved Pan-European indicators for sustainable forest management as adopted by the MCPFE Expert Level Meeting 7–8 October 2002, Vienna, Austria. http://www.mcpfe.org. Accessed Oct 2008.
Ministerial Conference on the Protection of Forests in Europe (MCPFE) (2007). Warsaw declaration. In Proceedings of the fifth ministerial conference on the protection of forests in Europe. Warsaw, 5–7 November 2007. http://www.mcpfe.org. Accessed Mar 2008.
Montes, F., Canellas, I., Alberdi, I., Condes, S., & Villanueva, J. A. (2005) Methodology used in the IFN, other forest inventories al regional scale and other plot nets in Spain for the assessment of the forest biodiversity through the dead wood, epiphytic lichens, stand structure and ground vegetation. Country report for forest focus- ComMon project.
Montrial Process, M. (1998). *Criteria and indicators for the conservation and sustainable management of temperate and boreal forests*. Montreal: The Montreal Process.
Montréal Process (2006). Criteria and indicators for the conservation and sustainable management of temperate and boreal forests. Montréal Process Liaison Office, International Forestry Cooperation Office, Forestry Agency, Japanese Ministry of Agriculture, Forestry and Fisheries, Tokyo, Japan. http://www.mpci.org/rep-pub/1995/santiago_e.html. Accessed Sept 2008.
Mountford, E. P. (2002). Fallen dead wood levels in the near-natural beech forest at La Tillaie reserve, Fontainbleau, France. *Forestry, 75*, 203–208.

Müller, J., Strätz, C., & Hothorn, T. (2005). Habitat factors for land snails in European beech forests with a special focus on coarse woody debris. *European Journal of Forest Research, 124*, 233–242.

Nilsson, S. G., Niklasson, M., Hedin, J., Aronsson, G., Gutowski, J. M., Linder, P., Ljunberg, H., Mikusiński, G., & Ranius, T. (2002). Densities of large living and dead trees in old-growth temperate and boreal forests. *Forest Ecology and Management, 161*, 189–204.

Nordén, B., & Appelqvist, T. (2001). Conceptual problems of ecological continuity and its bioindicators. *Biodiversity and Conservation, 10*, 779–791.

Norton, T. W. (1996). Conservation of biological diversity in temperate and boreal forest ecosystems. *Forest Ecology and Management, 85*, 1–7.

Noss, R. F. (1990). Indicators for monitoring biodiversity: a hierarchical approach. *Conservation Biology, 4*, 335–364.

Ohlson, M., Söderstrom, L., Hörnberg, G., Zackrisson, O., & Hermansson, J. (1997). Habitat qualities versus long-term continuity as determinants of biodiversity in boreal old-growth swamp forests. *Biological Conservation, 81*, 221–231.

Oosting, H. J. (1956). *The study of plant communities: an introduction to plant ecology*. San Francisco: WH Freeman and Co.

Palmer, S. C. F., Mitchel, R. J., Truscott, A.-M., & Welch, D. (2004). Regeneration failure in Atlantic oakwoods: the roles of ungulate grazing and invertebrates. *Forest Ecology and Management, 192*, 251–265.

Peterken, G. F. (1996). *Natural woodland: ecology and conservation in temperate regions*. Cambridge: Cambridge University Press.

Petriccione, B. (2006). Aspects of biological diversity in the CONECOFOR plots. VII. Naturalness and dynamical tendencies in plant communities. *Annali dell'Istituto Sperimentale per la Selvicoltura, 30*(Supplement 2), 93–96.

Pitkänen, S. (1997). Correlation between stand structure and ground vegetation: an analytical approach. *Plant Ecology, 131*, 109–126.

Pitkänen, S. (1998). The use of diversity indices to assess the diversity of vegetation in managed boreal forests. *Forest Ecology and Management, 112*, 121–137.

Poulsen, B. O. (2002). Avian richness and abundance in temperate Danish forests: tree variables important to birds and their conservation. *Biodiversity and Conservation, 11*, 1551–1566.

Pukkala, T. (2006). Optimising the semi-continuous cover forestry of Finland. *Allgemeine Forst- und Jagdzeitung, 177*(8/9), 141–149.

Rambo, T. R., & Muir, P. S. (1998a). Forest floor Bryophytes of Psedotsuga menziesii-Tsuga heterophylla stands in Oregon: influences of substrate and overstory. *Bryologist, 101*, 116–130.

Rambo, T. R., & Muir, P. S. (1998b). Bryophyte species associations with coarse woody debris and stand ages in Oregon. *Bryologist, 101*, 366–376.

Ratcliffe, P. R., & Peterken, G. F. (1995). The potential for biodiversity in British upland spruce forests. *Forest Ecology and Management, 79*, 153–160.

Roberts-Pichette, P., & Gillespie, L. (1999). Vegetation monitoring protocols working group of the biodiversity science board of Canada for the ecological monitoring and assessment network. EMAN occasional paper series report no. 9.EMAN Coordinating Office. Canada Centre for Inland Waters. P.O. Box 5050. Burlington, Ontario, Canada. L7R 4A6.

Rolston, H. (1990). Biology and philosophy in Yellowstone. *Biology and Philosophy, 5*, 241–258.

Rondeux, J. (1999). *La mesure des arbes et des eeoplemens foresteirs* (p. 522). Gembloux: Les Presses agronomiques de Bembloux.

Rondeux, J., & Sanchez, C. (2009). Review of indicators and field methods for monitoring biodiversity within National forest inventories. Core variable: Deadwood. Environmental monitoring and assessment. doi 10.1007/s10661-009-0917-6. Published online 05 May 2009. http://www.springerlink.com/content/3mt2512305u35g38/fulltext.pdf. Accessed Dec 2009.

Rondeux, J., Bertini, R., Bastrup-Birk, A., Corona, P., McRoberts, R. E., Sanchez, C., Ståhl, G., Winter, S., & Chirici. G. (submitted). Assessing deadwood using harmonised national forest inventory data. Forest Science.

Rondeux, J., Puissant, T., & Sanchez, C. (2005). Methodology used in the southern Belgium's forest, Inventory for the assessment of ground vegetation, deadwood and stand structure. Country report for forest focus – Common project. http://www.resgeom.slu.se/resana/projekt/common/documents.htm. Accessed Mar 2009.

Šaudytė, S., Karazihja, S., & Belova, O. (2005). An approach to assessment of naturalness for forest stands in Lithuania. *Baltic Forestry, 11*(1), 39–45.

Scherzinger, W. (1966). *Naturschutz im Wald* (p. 447). Stuttgart: Qualitätsziele einer dynamischen Waldentwicklung.

Schieler, K., Hauk, E. (2001). Instruktion für die Feldarbeit, Österreichische Waldinventur 2000/2002 Dienstanweisung FBVA, 99–101. http://web.bfw.ac.at/700/pdf/da_ges_neu.pdf. Accessed Mar 2006.

Scholes, R. J., & Biggs, R. (2005). A biodiversity intactness index. *Nature, 434*, 45–49.

Schulte, L. A., Mitchell, R. J., Hunter, M. L., Franklin, J. F., McIntyre, R. K., & Palik, B. J. (2006). Evaluating the conceptual tools for forest biodiversity conservation and their implementation in the U.S. *Forest Ecology and Management, 232*, 1–3.

Schulze, E. D., & Mooney, H. A. (1994). Ecosystem function of biodiversity: a summary. In E. D. Schulze & H. A. Mooney (Eds.), *Biodiversity and ecosystem function* (pp. 497–510). Berlin: Springer-Verlag.

Schumacher, H. (2006). *Zum Einfluss forstlicher Bewirtschaftung auf die Avifauna von Rotbuchenwäldern im nordostdeutschen Tiefland.* Göttingen: Cuvillier Verlag.

Siipilehto, J., & Siitonen, J. (2004). Degree of previous cutting in explaining the differences in diameter distributions between mature managed and natural Norway spruce forests. *Silva Fennica, 38*(4), 425–435.

Siitonen, J. (2001). Forest management, coaose woody debris and saproxylic organisms: Fennoscandian boreal forest as an example. *Ecological Bulletin, 49*, 11–41.

Smith, P. G. R., & Theberge, J. B. (1987). Evaluating natural areas using multiple criteria: Theory and practice. *Environmental Management, 11*(4), 447–460.

Smith, G., Gittings, T., Wilson, M., French, L., Oxbrough, A., O'Donoghue, S., Pithon, J., O'Donnell, V., McKee, A., Iremonger, S., O'Halloran, J., Kelly, D., Mitchell, F., Giller, P., & Kelly, M. (2005). BIOFOREST. Assessment of biodiversity at different stages of the forest cycle. Final report, Feb 2005.

Soulé, M. E. (1985). What is conservation biology? *Bioscience, 35*(11), 727–734.

Stokland, J. N., Tomter, S. M., & Söderberg, U. (2004). Development of deadwood indicators for biodiversity monitoring: experiences from Scandinavia. In: Monitoring and indicators of forest biodiversity in Europe-From ideas to operationality. *European Forest Institute Proceedings, 51*, 207–226.

Sukopp, H., Hejný, S., & Kovarik, I. (Eds.). (1990). *Urban ecology. Plants and plant communities in urban environments.* The Hague: SPB Academic Publications.

Tardif, J., & Bergeron, Y. (1999). Population dynamics of *Fraxinus nigra* in response to flood-level variations in North-western Quebec. *Ecological Monographs, 69*(1), 107–125.

Thimonier, A., Keller, W., Dupouey, J. L. (2003). Nitrogen and ground vegetation. http://www.wsl.ch/forest/risks/projects/nitrogen/. Accessed Mar 2008.

Trass, H., Vellak, K., & Ingerpuu, N. (1999). Floristical and ecological properties for identifying of primeval forests in Estonia. *Annales Botanici Fennici, 36*, 67–80.

United Nations Economic Commission for Europe and United Nations Food and Agriculture Organization (UN/ECE-FAO) (2000). Forest resources of Europe, CIS, North America, Australia, Japan and New Zealand: Main report. United Nations Publication GE.00-21547. Geneva, Switzerland. pp. 443.

United Nations Environment Programme (UNEP) (2001). Indicators and environmental impact assessment: Designing national-level monitoring and indicator programmes, UNEP/CBD/SBSTTA/7/12, Subsidiary body on scientific, technical and technological advice. http://www.biodiv.org/doc/meetings/sbstta/sbstta-07/official/sbstta-07-12-en.pdf. Accessed Mar 2009.

Uotila, A., Kouki, J., Kontkanen, H., & Pulkkinen, P. (2002). Assessing the naturalness of boreal forests in eastern Fennoscandia. *Forest Ecology and Management, 161*, 257–277.

Usher, M. B. (1986). Wildlife conservation evaluation: attributes, criteria and values. In M. D. Usher (Ed.), *Wildlife conservation evaluation: attributes, criteria and values* (pp. 1–69). Cambridge: Chapman and Hall.

von Oheimb, G., Westphal, C., Tempel, H., & Härdtle, W. (2005). Structural pattern of a near-natural beech forest (*Fagus sylvatica*) (Serrahn, North-east Germany). *Forest Ecology and Management, 212*, 253–263.

Westphal, C., Tremer, N., von Oheimb, G., Hansen, J., von Gadow, H., & Härdtle, W. (2006). Is the reverse J-shaped distribution universally applicable in European virgin beech forests? *Forest Ecology and Management, 223*, 75–83.

Wilson, S. M., Pyatt, D. G., Malcolm, D. C., & Connolly, T. (2001). The use of ground vegetation and humus type as indicators of soil nutrient regime for an ecological site classification of British forests. *Forest Ecology and Management, 140*(2/3), 101–116.

Winter, S. (2006). Naturnähe-Indikatoren für Tiefland-Buchenwälder. Indicators for naturalness of lowland beech forests. *Forstarchiv, 77*(3), 94–101.

Winter, S., & Möller, G. C. (2008). Microhabitats in lowland beech forests as monitoring tool for nature conservation. *Forest Ecology and Management, 255*, 1251–1261.

Winter, S., Flade, M., Schumacher, H., Kerstan, E., & Möller, G. (2005). The importance of near-natural stand structures for the biocoenosis of lowland beech forests. *Forest, Snow and Landscape Research, 79*(1/2), 127–144.

Woodley, J., Kay, J., & Francis, G. (Eds.). (1993). *Ecological integrity and the management of ecosystems*. Delray Beach: St Lucie Press.

Willoughby, I., Jinks, R., Gosling, P., & Kerr, G. (2004). *Creating new broadleaved woodland by direct seeding. Forestry commission practice guide* (p. 32). Edinburgh: Forestry Commission.

World Wildlife Fund (WWF) (2004). Deadwood – Living forest. The importance of veteran trees and deadwood to biodiversity. WWF report. Oct 2004. pp. 19.

Zerbe, S., & Kempa, D. (2005). Waldumbau und Biodiversität. Unterschiedliche forstliche Entwicklungsziele und deren Auswirkungen auf die biologische Vielfalt. *Naturschutz und Landschaftsplanung, 37*(4), 106–114.

Chapter 4
The Common NFI Database

Gherardo Chirici, Roberta Bertini, Davide Travaglini, Nicola Puletti, and Ugo Chiavetta

Abstract To test bridging techniques for the harmonized estimation of forest biodiversity indicators for each of the selected essential features a common database was constructed and populated with raw NFI data provided by some of the COST Action E43 participating countries. The database was structured with five tables in a relational database: one table for descriptive plot data, one for tree level data, one for deadwood pieces, one for shrub data and one for ground vegetation. The database was populated with data for 320,023 trees, data for 25,639 pieces of deadwood, 12,588 shrub records and 34,364 ground vegetation records from 14,638 NFI plots provided by 13 European countries and the USA.

4.1 Construction of the Common NFI Database

A large number of bridging techniques based on a wide range of theoretical considerations could be used to harmonize estimates of biodiversity indicators calculated using NFI data. Although all bridging techniques may have advantages and disadvantages, they all may be difficult to evaluate without empirical tests. In fact, because harmonization processes are frequently quite innovative, it may be that technical solutions are seldom tested. An experimental test phase is essential to support the selection of bridging techniques and to evaluate the effects of different reference definitions and bridging options on the overall success of the harmonization process. To support and facilitate this test phase, WG3 of COST Action E43 constructed a common relational database (DB) populated with raw data voluntarily

G. Chirici (✉) and U. Chiavetta
Università degli Studi del Molise, Italy
e-mail: gherardo.chirici@unimol.it; uchiavetta@unimol.it

R. Bertini, D. Travaglini, and N. Puletti
Università degli Studi di Firenze, Italy
e-mail: bertini.roberta@gmail.com; davide.travaglini@unifi.it; nicola.puletti@unifi.it

G. Chirici et al. (eds.), *National Forest Inventories: Contributions to Forest Biodiversity Assessments*, Managing Forest Ecosystems 20,
DOI 10.1007/978-94-007-0482-4_4,

contributed by the NFIs of some participating countries. The purpose of the DB was to facilitate assessment of the technical feasibility and effects of different harmonization options, not to compare estimates of biodiversity indicators. The effort consisted of two steps: (1) construct a common DB structure and (2) populate the common DB with current data from individual NFI databases.

The DB was populated with data for the seven essential biodiversity features: (Sect. 3.2) forest categories, forest age, forest structure, deadwood, regeneration, ground vegetation and natural forests. All the essential features were selected for their suitability for forest biodiversity assessment as described in the previous phases of the harmonization process (Chap. 2; Winter et al. 2008). The relational DB was organized in five tables corresponding to the nature of the NFI data:

- PLOT: A table consisting of 15 fields with each record representing a single NFI plot, the basic sampling unit used by all NFIs participating in COST Action E43.
- TREE: A table consisting of 16 fields with each record representing a single observed tree on a single NFI plot. Data relate to "*stems*" not to "*trees*," but we prefer to use the less precise but more common word "*trees*". Definitions of both trees and stems may be found in the COST Action E43 reference definitions (Vidal et al. 2008; Gschwantner et al. 2009; Lanz et al. 2009; Tomppo et al. 2010).
- DEADWOOD: A table consisting of 14 fields with each record representing a single deadwood piece observed on a single NFI plot. For WG3 assessments, deadwood included lying and standing deadwood but not stumps, because only a few NFIs collect stump data.
- SHRUB: A table consisting of five fields with each record representing the shrub coverage for a single NFI plot.
- GROUND VEGETATION: A table consisting of five fields with each record representing the ground vegetation coverage for a single NFI plot.

The DB is completed by a Look-Up table with six fields containing the definitions of the codes used in the other five tables. The fields in the five thematic tables of the common NFI DB are described in Tables 4.1–4.6.

4.2 Population of the Common NFI Database

The DB structure was distributed to the NFIs of all countries participating in COST Action E43 with a request to populate it with a selection of their raw NFI data. Despite construction of the DB to facilitate storage of NFI data from different sources, modifications of the data for some countries were necessary. For example, circumference measures of trees needed to be converted to diameter at-breast-height (dbh), and measures in English imperial units needed to be converted to metric units. All the plots from European NFIs were classified (PLOT, field forest category) according to the 14 European forest categories based on the

Table 4.1 Structure of the PLOT table

Field	Description
Country	Country code
Plot id	Plot number, unique within a country
Forest category	Numeric code from the European Forest Type (EEA 2006) system of nomenclature (*category* level) for European plots. Original FIA codes for plots from the USA
Naturalness	Information based on the country classification acquired in the field. All information based on local systems of nomenclature potentially useful for evaluating plot naturalness (e.g.: the relationship between real and potential vegetation and/or the absence of anthropogenic disturbances) were included
Origin	Information based on the country system of nomenclature dealing with the prevalent origin of trees within the plot (natural regeneration, seeding, planting, etc.)
Cutting system	Information based on the country system of nomenclature dealing with the prevalent cutting system adopted within the plot (clearcutting, thinning, natural evolution, etc.)
Years since last treatment	Number of years elapsed since the last silvicultural treatment
Other human activities	Information based on the country system of nomenclature dealing with prevalent human activities other than thinning or forest cutting within the plot (grazing, soil preparation, fertilizing, roads, etc.)
Forest age	Information based on the country system of nomenclature dealing with the prevalent forest age/development stage of the plot as it is assessed in the field
Layers	Number of vertical layers (e.g.: one, two, many) based on the country system of nomenclature as it is assessed in the field
Regeneration cover	Percentage or cover classes of the trees regenerated from seeds based on the country system of nomenclature as it is assessed in the field
Regeneration number	Number of individuals (or in classes) per hectare of the trees regenerated from seeds based on the country system of nomenclature as it is assessed in the field
Regeneration number capable of further development	Number of individual trees (or in classes) per hectare regenerated from seeds judged to be capable of further development based on the country system of nomenclature as it is assessed in the field
Crown cover	Crown cover (in cover classes or percentages) as it is assessed in the field
Stage of development	Stage of development (e.g.: seedling stage, thinning stage) based on the country system of nomenclature as it is assessed in the field

nomenclature developed by EEA (2006), (Sect. 5.1). Additionally we used the ICP code (table TREE, field ICP code) for the tree species. The forest categories (EEA 2006) system is available for European plots only; the plots from the USA were classified according to the system of nomenclature based on the dominant tree species used by the Forest Inventory and Analysis programme of the U.S. Forest Service.

Table 4.2 Structure of the TREE table

Field	Description
Country	Country code
Plot id	Plot number, unique within a country, consistent with the same field in the PLOT table
Tree id	Tree number, unique within the plot
Genus	Tree genus, according to the scientific system of nomenclature (Flora Europaea)
Species	Tree species, according to the scientific system of nomenclature (Flora Europaea)
ICP code	Genus + species code based on the ICP code system (ICP Forests 2009)
Nativeness	Nativeness of the species in its local natural range. It could be expressed by the classes "site-native" if the tree species is in its local natural range, "introduced" if the tree species is out of its natural local range, or with other country systems of nomenclature
dbh	Diameter at breast height (mm), One for each tree
Sampling unit area	Sampling area of the plot on which the tree is observed or measured (m^2)
Basal area factor	In the case of relascopic areas, the Bitterlich factor used (m^2 ha^{-1})
Measured height	As it is measured or visually estimated in the field (m)
Modelled height	As estimated from models, usually based on diameter and species (m)
Age	Tree age in years
Age method	Method used for calculating or estimating tree age in the country
Management system	Information based on the country system of nomenclature dealing with the prevalent management system adopted (high forest, coppice, etc.)
Social position	Social position of the tree such as dominant, co-dominant, or in other classification systems

In general, the resulting DB was quite heterogeneous. Data for the same field of the same table were provided by some countries in categories whereas other countries provided data using values of continuous variables. In addition, practical meanings of data provided for the same field and table differed for different countries. In some cases the DB structure was too simple or too complex for the information available in the country NFI DBs. As a result, the raw data from individual NFI DBs sometimes needed to be aggregated and in other cases needed to be disaggregated. For example, aggregation was needed for data for the regeneration cover field of the PLOT table because some countries assess regeneration by species. In addition, aggregation was necessary in some cases for data for the other human activities field of the PLOT table because some countries use more complex information structures involving several fields. Finally, not all the information requested for the common DB is assessed by all the countries; consequently some fields for some countries are empty. Two different codes were used to avoid empty cells in the DB: "data not assessed" was used when a country does not assess a variable in the field, while "data not available" was used when a country assesses a variable but the information for a specific sampling unit was missing. The latter situation could frequently be attributed to assessment of a variable for only a sub-sample of the sampling units in the population. The digit '0'

Table 4.3 Structure of the DEADWOOD table

Field	Description
Country	Country code
Plot id	Plot number, unique within a country, consistent with the same field in the PLOT table
Piece id	Number of each single piece of deadwood, unique within each plot
Lying standing	Position of the piece of deadwood: "L" for lying and "S" for standing (the threshold is 45° angle with the vertical position); or classification using country systems
Inventorying method	LIS for Line Intersect Sampling, PLOT for plot sampling
Plot area	Area of the plot where the deadwood is assessed (m^2); used only if inventorying method is PLOT
LIS length	Length of the line transect (m); used only if inventorying method is LIS
Diameter 1	For lying deadwood when inventorying method is LIS: diameter of the deadwood piece at the intersection of the line transect (mm)
	For lying deadwood when inventorying method is PLOT: the diameter at half length of the deadwood piece if it is assessed by the method usually called "median diameter" or the median diameter or the first diameter (minimum or smallest) of the two diameters for the average diameter method (mm)
	For standing deadwood: the diameter usually used by the country for calculating the standing deadwood (it could be the dbh for standing dead trees, a median diameter or the first diameter for the average diameter method) (mm)
Diameter 2	Following the concept of diameter 1 for lying or standing deadwood if the average diameter method is used this is the second diameter (maximum or larger) (mm)
Length	For standing deadwood: height of the tree or the height of the stem (m)
	For lying deadwood: length of the deadwood piece (m). In the case of the average diameter method this is the distance between the points where diameter 1 and diameter 2 were measured
Volume model	The mathematical model used to estimate the volume on the basis of diameter(s) and length
Decay	Decay stage according to the country classification system
Forest categories	The forest types or category (e.g.: broadleaves, coniferous)
Volume	Deadwood piece volume (it can refer to single deadwood pieces or dead trees, lying or standing)

Table 4.4 Structure of the SHRUB table

Field	Description
Country	Country code
Plot id	Plot number, unique within a country, consistent with the same field in the PLOT table
Species	Numeric code according to the ICP system of nomenclature for the shrub species (ICP Forests 2009)
Cover	Percentage cover of the species within the plot; within a plot, the sum of cover values may be less or more than 100%
Date	Date of the field assessment, or other information on the month or season if the date is not available

Table 4.5 Structure of the GROUND VEGETATION table

Field	Description
Country	Country code
Plot id	Plot number, unique within a country, consistent with the same field in the PLOT table
Groups or species	Groups of species or species according to the country system of nomenclature
Cover	Percentage cover of the species within the plot; within a plot, the sum of cover values may be less or more than 100%
Date	Date of the assessment in the field, or other information on the month or season if the date is not available

Table 4.6 Structure of the Look Up table

Field	Description
DB table	Name of the DB table to which the variable is related
Variable	Name of the field (e.g., *Social position, Decay*) to which the description refers
Country	Unique country code using abbreviations
Code	Code value of the class according to the adopted country system of nomenclature
Description	Short description of the code (e.g., co-dominant; decayed)
Note	Long description or comment on the code if needed

was used when a variable was assessed in the field but the quantitative result of the assessment was zero, for example when no deadwood was found on a plot.

Although construction and population of the DB was considered a pre-harmonization stage, the very process of using raw, national NFI data in a variety of original formats to populate the common DB infrastructure constituted a first harmonization step.

4.3 Characteristics of the Common NFI Database

The NFIs of 13 European countries and the USA populated the common DB with data for 14,638 plots selected from their NFI DBs (Figs. 4.1 and 4.2, Table 4.7).

Austria, Switzerland and Germany selected plots for populating the common DB on the basis of a systematic or a random sample of all plots, whereas all other countries selected plots for one or more geographic regions (Fig. 4.1, Table 4.7).

All countries provided data for the PLOT and TREE tables; ten countries provided data for the DEADWOOD table; and nine and eight countries provided data for the SHRUB and GROUND VEGETATION tables, respectively. Only Switzerland, the Czech Republic, Germany, Denmark, Spain and Italy provided information for all five DB tables (Table 4.8).

The data used to populate the five tables of the common DB are briefly described in the following sections. For some of the countries the plot information was acquired within the framework of local forest inventories which are not formally

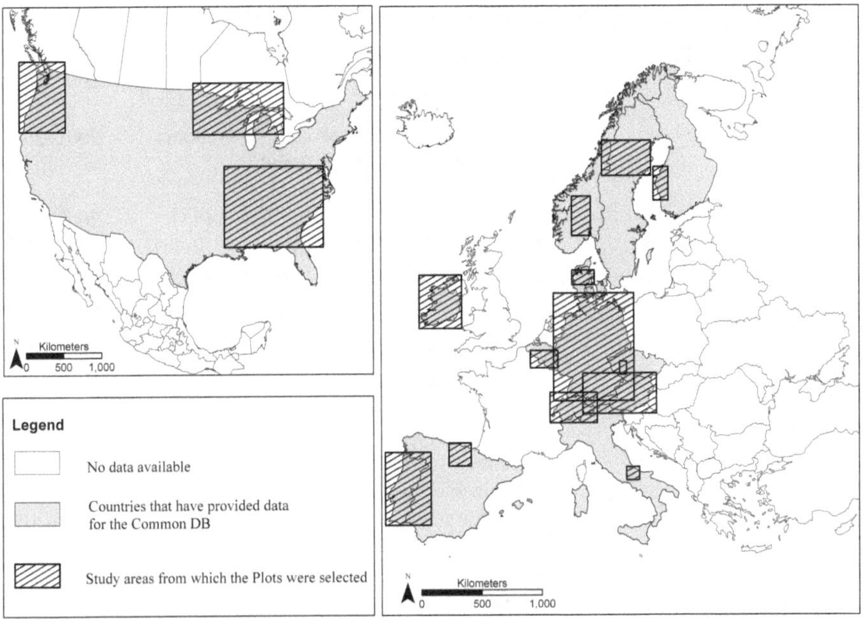

Fig. 4.1 The 14 countries whose NFIs contributed data (grey) and their related geographic area (squares). The ecoregions in the USA are 242 in the far west, 212 in the north central, and 231 in the southeast

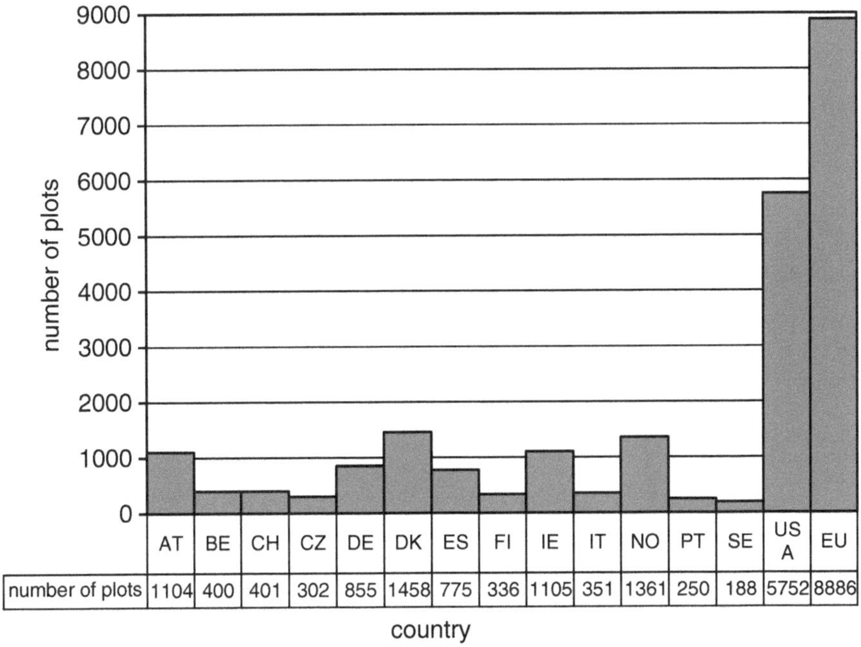

Fig. 4.2 Number of plots provided by countries. EU denotes the aggregation of all European countries contributing data

Table 4.7 Data selection method and area covered (see Fig. 4.1)

Country	Number of plots	Selection method	Region	Years
Austria	1,104	Random sampling of 10% for each forest type	Austria	2000–2002
Belgium	400	Selection of the plots for four forest types: 6.4; 5.1; 4.2 and 14.2 (EEA 2006) in Wallonia	Wallonia	2001–2006
Czech Republic	302	All "forest" plots included in selected forest enterprises in Středočeský kraj	Central Bohemia	2005
Denmark	1,458	All plots included in the administrative region Midt Jutland in the first cycle of the Danish NFI 2002–2006	Midt Jutland	2002–2006
Finland	336	All plots on forest land (FRA definition) in the Forestry Centre of Rannikko, Pohjanmaa measured in 2006	Rannikko	2006
Germany	855	Systematic selection from the original national dataset on the basis of a 16-km × 16-km grid	Germany	2001–2002
Ireland	1,105	Subsample of the national dataset	Ireland	–
Italy	351	Local forest inventory in Regione Molise. Same methods and definitions of NFI	Molise	2005–2009
Norway	1,361	All the plots for Oppland county	Oppland	–
Portugal	250	Subsample of the national dataset	Portugal	2005–2008
Spain	775	All plots in the administrative region of Alava	Alava	1997–2007
Sweden	188	All temporary plots on forestland measured in the administrative region "The county of Västerbotten"	Västerbotten	2003
Switzerland	401	Selection of 100 plots for four of the most common forest types with an altitude nearest the mean altitude characteristics of each forest type	Switzerland	1993–1995
USA	5,752	All plots in five selected forest types in the selected ecological provinces (ecoregions 212, 231, 242)	Ecoregions 212, 231, 242	1998–2002

"national". For simplicity we characterize all data in the dataset as coming from NFIs even if they were sometimes acquired and provided by local subnational administrations (e.g.: Wallonia in Belgium or Molise in Italy).

Table 4.8 Countries contributing data to the common DB tables

Country	Plot	Tree	Deadwood	Shrub	Ground Vegetation	Total
Austria	1	1	0	1	0	3
Belgium	1	1	1	0	1	4
Czech Republic	1	1	1	1	1	5
Denmark	1	1	1	1	1	5
Finland	1	1	1	0	0	3
Germany	1	1	1	1	1	5
Ireland	1	1	0	0	0	2
Italy	1	1	1	1	1	5
Norway	1	1	0	0	0	2
Portugal	1	1	0	1	1	4
Spain	1	1	1	1	1	5
Sweden	1	1	1	1	0	4
Switzerland	1	1	1	1	1	5
USA	1	1	1	0	0	3
Total	14	14	10	9	8	–

1 = available, 0 = not available

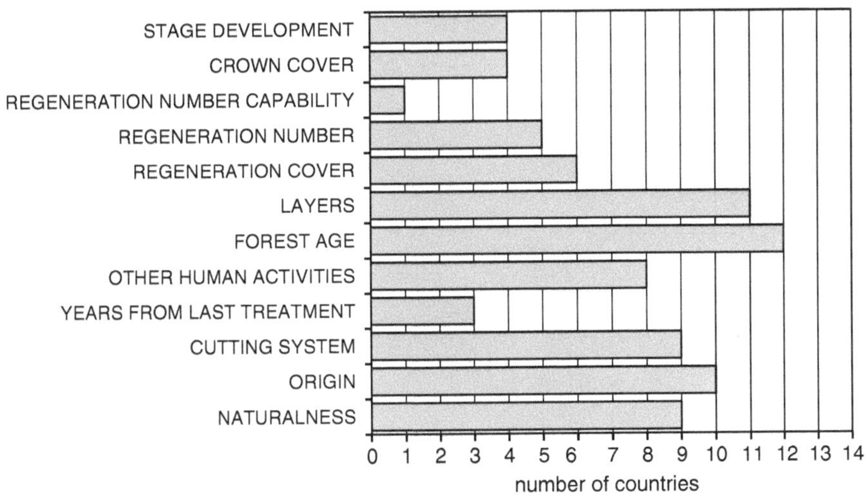

Fig. 4.3 Available data for the fields of the PLOT table ("not assessed" data are not considered as provided by the countries)

4.3.1 PLOT Table

No country contributed data for all 14 fields of the PLOT table. However, all except two countries provided data for the Forest Age field, the most frequently populated field for this table, and 11 countries provided data for the Layers field. The NFIs of Sweden and Switzerland provided data for the largest number of fields, 11 and 10 fields, respectively. A complete overview of the data available for each field of the PLOT table is reported in Fig. 4.3 and Table 4.9.

Table 4.9 Data available (1) and data not available (0) per country and per field of the PLOT table

Variable	Austria	Belgium	Czech Republic	Denmark	Germany	Finland	Ireland	Italy	Norway	Portugal	Spain	Sweden	Switzerland	USA	Total
Forest category	1	1	1	1	1	1	1	1	1	1	1	1	1	1	14
Naturalness	0	1	1	1	1	1	0	1	0	0	1	1	1	0	9
Origin	0	1	0	1	1	1	0	1	0	1	1	1	1	1	10
Cutting system	0	0	0	1	1	1	0	1	1	1	1	1	1	0	9
Years from last treatment	0	0	0	1	0	0	0	0	0	0	0	0	1	1	3
Other human activities	0	0	0	1	0	1	0	1	0	1	1	1	1	1	8
Forest age	1	1	1	1	1	1	0	1	1	0	1	1	1	1	12
Layers	1	0	1	1	1	1	0	1	1	0	1	1	1	1	11
Regeneration cover	1	0	1	0	1	0	0	0	0	1	0	1	1	0	6
Regeneration number	0	0	1	0	0	1	0	0	1	0	0	1	0	1	5
Regeneration number capability	0	0	0	0	0	1	0	0	0	0	0	0	0	0	1
Crown cover	0	0	0	0	0	0	0	1	0	0	1	1	1	0	4
Stage of development	0	0	0	0	0	0	0	1	0	0	1	1	1	0	4
Total	4	4	6	8	7	9	1	9	5	5	9	12	11	7	–

0 = not assessed

4.3.2 TREE Table

The TREE table was populated with data for 320,023 trees, or more precisely stems of trees (Vidal et al. 2008) (Fig. 4.4). The USA contributed data for 179,087 trees representing more than 140 species with *Populus tremuloides*, *Pseudotsuga menziesii*, *Abies balsamea*, *Betula papyrifera*, *Acer ubrum* (representing 62% of the total) the most common. Data for the remaining 140,936 trees were contributed by the NFIs of European countries and represent more than 130 species of which *Picea abies*, *Picea sitchensis*, *Fagus sylvatica*, *Betula pubescens*, *Pinus sylvestris* are the five most common and represent 56% of the total; 27% of the trees from the NFIs of European countries are *Picea abies*. The largest diameter tree was a *Pseudotsuga menziesii* from Ecoregion 242 in the USA with dbh of 2,849 mm and height of 90 m. The largest diameter European tree was an *Aesculus hippocastanum* from Ireland with dbh of 1,447 mm, while the tallest European tree was a *Picea abies* of 50 m from Switzerland.

All 14 countries provided data for tree *Genus*, *Species* and dbh. A complete overview of the data available for each field of the TREE table by country is reported in Fig. 4.5 and Table 4.10. The unique international identification codes for species for the *ICP code* field were entered by the DB administrators after data submission. This action is a pre-processing, harmonization step. The NFIs of the Czech Republic and Denmark contributed data for the largest number of fields of the TREE table.

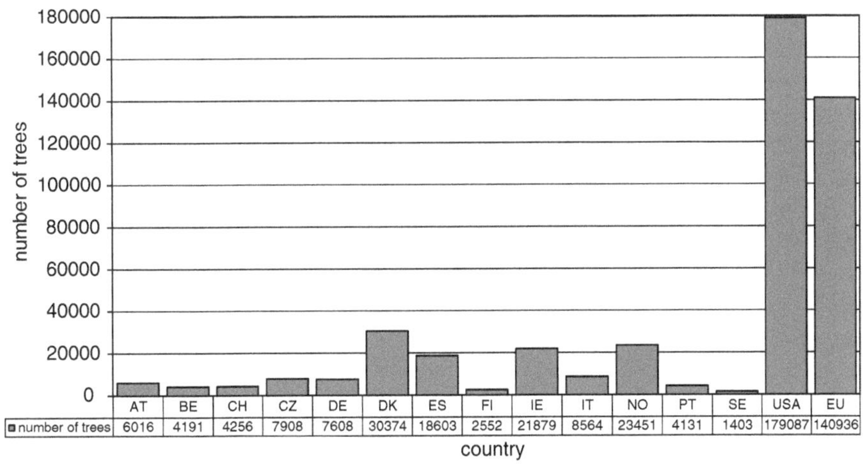

Fig. 4.4 Trees per country based on TREE table of the common NFI DB. EU denotes the aggregation of all European countries contributing data

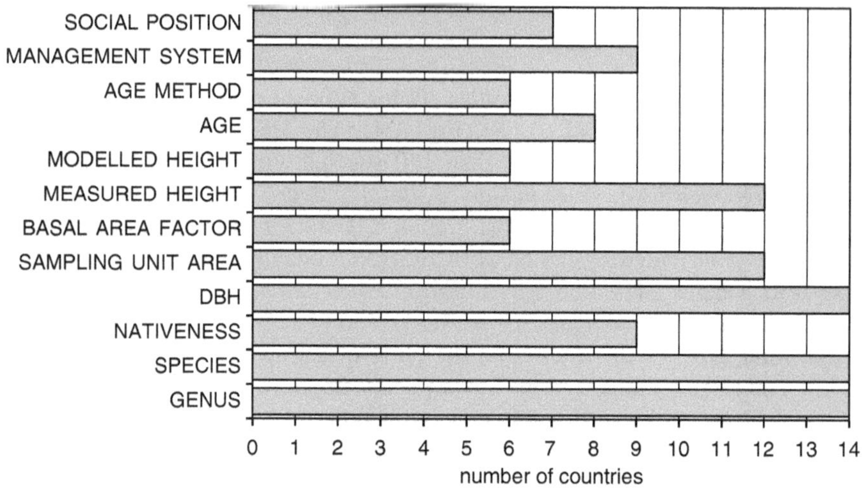

Fig. 4.5 Available data for the fields of the PLOT table ("not assessed" data are not considered as provided by the countries)

4.3.3 DEADWOOD Table

The DEADWOOD table was populated by the NFIs of ten countries with data for 25,639 pieces of deadwood, approximately half of them from the USA (Fig. 4.6). Almost all ten countries populated most of the fields of the table. The only fields of the DEADWOOD table for which only a limited number of countries submitted data were *Diameter 2*, *Volume Function* and *LIS/Length*. These limitations are attributed to specific deadwood sampling methods or field measurement approaches. As shown in Fig. 4.7, the number of lying deadwood pieces is greater than the number of standing dead trees for eight of ten countries. A complete overview of the data available for each field of the DEADWOOD table by country is reported in Fig. 4.8 and Table 4.11.

The coarsest piece of lying deadwood was from a German broadleaf species tree and had a median diameter of 930 mm. The longest piece of lying deadwood pieces was from a broadleaf species tree from the USA and had length of 73 m; for European countries, both Germany and the Czech Republic assessed a piece of deadwood of length 30 m. The standing dead tree with the largest diameter was a European coniferous tree with dbh of 1,040 mm of dbh, while the tallest dead tree was from a broadleaf species in the USA with height of 34 m.

Table 4.10 Available data by field and country for the TREE table

Variable	Austria	Belgium	Czech Republic	Denmark	Germany	Finland	Ireland	Italy	Norway	Portugal	Spain	Sweden	Switzerland	USA	Total
Genus	1	1	1	1	1	1	1	1	1	1	1	1	1	1	14
Species	1	1	1	1	1	1	1	1	1	1	1	1	1	1	14
ICP Code	0	0	0	0	0	0	0	0	0	0	0	0	0	0	0
Nativeness	0	1	1	0	0	1	0	1	1	1	0	1	1	1	9
dbh	1	1	1	1	1	1	1	1	1	1	1	1	1	1	14
Sampling unit area	1	1	1	1	0	0	1	1	1	1	1	1	1	1	12
Basal area factor	1	0	0	1	1	1	0	0	0	0	1	0	1	0	6
Measured height	1	1	1	1	0	1	0	1	1	1	1	1	1	1	12
Modelled height	1	0	1	1	1	0	1	0	0	0	0	0	1	0	6
Age	0	0	1	1	1	1	1	1	0	1	0	1	0	0	8
Age method	0	0	1	1	0	1	1	1	0	0	0	1	0	0	6
Management system	0	1	1	1	0	0	0	1	1	1	1	1	1	0	9
Social position	0	0	1	1	0	1	0	1	0	0	0	1	1	1	7
Total	7	7	11	11	6	9	7	10	7	8	7	10	10	7	–

"0" indicates not available and includes "Not assessed" and "1" indicates available

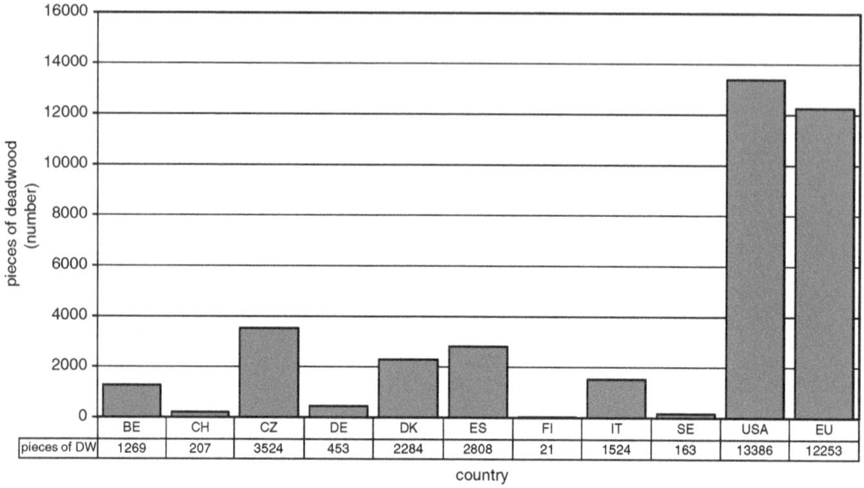

	BE	CH	CZ	DE	DK	ES	FI	IT	SE	USA	EU
pieces of DW	1269	207	3524	453	2284	2808	21	1524	163	13386	12253

Fig. 4.6 Number of pieces of deadwood by country from the DEADWOOD table. EU denotes the aggregation of all European countries contributing data

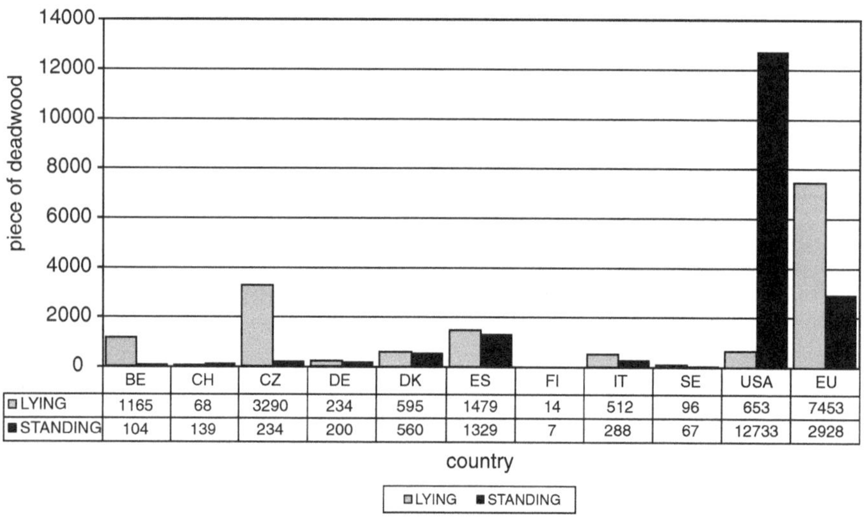

	BE	CH	CZ	DE	DK	ES	FI	IT	SE	USA	EU
LYING	1165	68	3290	234	595	1479	14	512	96	653	7453
STANDING	104	139	234	200	560	1329	7	288	67	12733	2928

Fig. 4.7 Distribution of the lying and standing deadwood by country. EU denotes the aggregation of all European countries contributing data

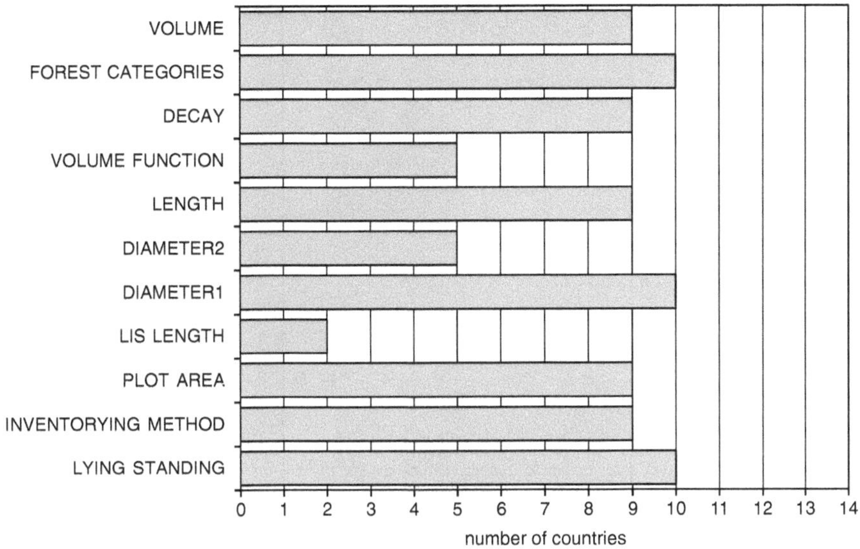

Fig. 4.8 Data availability in the fields of the table DEADWOOD ("not assessed" data are not considered as provided by the countries)

4.3.4 *SHRUB Table*

The SHRUB table was populated with 12,588 records provided by nine European countries; Spain, Austria and Denmark provided most of the data (Fig. 4.9). Nearly all nine countries provided data for most of the fields of this table. A complete overview of the data available for each field of the SHRUB table is reported in Fig. 4.10 and Table 4.12.

4.3.5 *GROUND VEGETATION Table*

Eight European countries submitted data for 34,364 records for the GROUND VEGETATION table (Fig. 4.11). All eight countries submitted data for the Groups or Species, Cover and Date fields. A complete overview of the data available for the fields of the SHRUB table is reported in Fig. 4.12 and Table 4.13.

Table 4.11 Available data by field and country for the DEADWOOD table

Variable	Austria	Belgium	Czech Republic	Denmark	Finland	Germany	Ireland	Italy	Norway	Portugal	Spain	Sweden	Switzerland	USA	Total
Lying/Standing	0	1	1	1	1	1	0	1	0	0	1	1	1	1	10
Inventory method	0	1	1	1	1	1	0	1	0	0	0	1	1	1	9
Plot area	0	1	1	1	1	1	0	1	0	0	1	1	1	0	9
LIS length	0	0	0	0	0	0	0	1	0	0	0	0	0	1	2
Diameter 1	0	1	1	1	1	1	0	1	0	0	1	1	1	1	10
Diameter 2	0	0	0	0	1	0	0	1	0	0	1	1	0	1	5
Length	0	1	1	1	1	1	0	1	0	0	1	1	0	1	9
Volume function	0	1	1	1	0	0	0	0	0	0	1	0	0	1	5
Decay	0	1	1	1	1	1	0	1	0	0	1	1	0	1	9
Forest category	0	1	1	1	1	1	0	1	0	0	1	1	1	1	10
Volume	0	1	1	0	1	1	0	1	0	0	1	1	1	1	9
Total	0	9	9	8	9	8	0	10	0	0	9	9	6	10	–

"0" indicates not available and includes "not assessed" and "1" indicates available

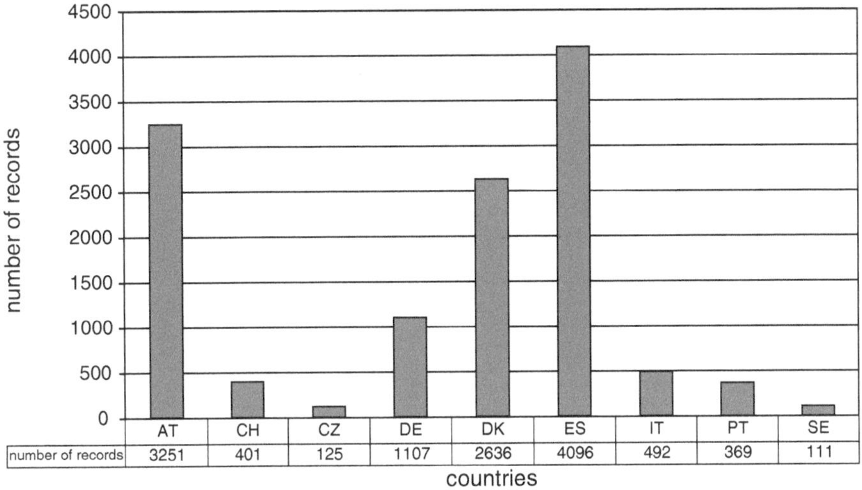

Fig. 4.9 Number of data records by country for the SHRUB table

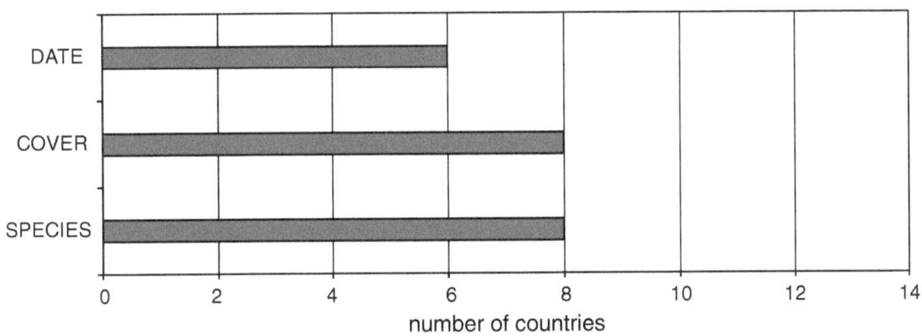

Fig. 4.10 Available data for the fields of the SHRUB table ("not assessed" data are not considered as data provided by the countries)

Table 4.12 Available data by field and country for the GROUND VEGETATION table

Variable	Austria	Belgium	Czech Republic	Denmark	Finland	Germany	Ireland	Italy	Norway	Portugal	Spain	Sweden	Switzerland	USA	Total
Species	1	0	1	1	0	1	0	1	0	1	1	1	0	0	8
Cover	1	0	1	1	0	1	0	0	0	1	1	1	1	0	8
Date	1	0	1	0	0	1	0	0	0	1	1	0	1	0	6
Total	3	0	3	2	0	3	0	1	0	3	3	2	2	0	–

"0" indicates not available and includes "not assessed" and "1" indicates available

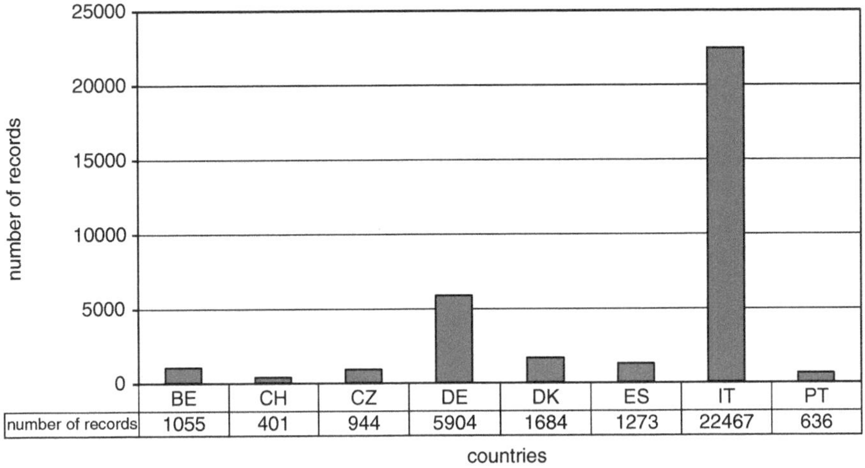

Fig. 4.11 Records provided by country for the GROUND VEGETATION table

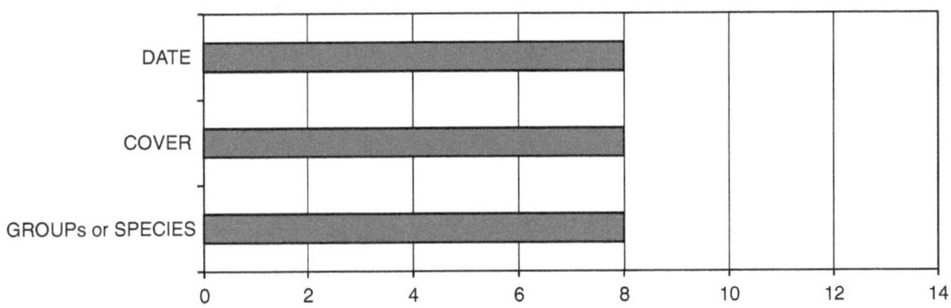

Fig. 4.12 Available data for the fields of the GROUND VEGETATION table by country ("not assessed" data are not considered as provided by the countries)

Table 4.13 Available data by field and country for the SHRUB table

Variable	Austria	Belgium	Czech Republic	Denmark	Finland	Germany	Ireland	Italy	Norway	Portugal	Spain	Sweden	Switzerland	USA	Total
Groups or Species	0	1	1	1	0	1	0	1	0	1	1	0	1	0	8
Cover	0	1	1	1	0	1	0	1	0	1	1	0	1	0	8
Date	0	1	1	1	0	1	0	1	0	1	1	0	1	0	8
Total	0	3	3	3	0	3	0	3	0	3	3	0	3	0	–

"0" indicates not available and includes "Not assessed" and "1" indicates available

References

EEA (European Environmental Agency). (2006). European forest types – Categories and types for sustainable forest management reporting and policy. EEA Report 9. Office for Official Publications of the Communities, Luxembourg. 111 pp.

Gschwantner, T., Schadauer, K., Vidal, C., Lanz, A., Tomppo, E., di Cosmo, L., Robert, N., Englert Duursma, D., & Lawrence, M. (2009). Common tree definitions for national forest inventories in Europe. *Silva Fennica, 43* (2), 303–321.

ICP Forests (International Co-operative Programme on Assessment and Monitoring of Air Pollution Effects on Forests). (2009). Code lists vascular plants. Version Nov 2009. Expert panel on ground vegetation and biodiversity. Available on-line at http://www.icp-forests.org/EPbiodiv.htm. Last accessed Dec 2009.

Lanz, A., Alberdi, I., Barbati, A., Barsoum, N., Brändli, U-B., Chirici, G., Cienciala, E., Di Cosmo, L., Freudenschuss, A., Gabler, K., Gschwantner, T., Hylen, G., Ilvesniemi, H., Kusar, G., Kändler, G., Lawrence, M., McRoberts, R. E., Nabuurs, G-. J., Petersson, H., Priwitzer, T., Robert, N., Rondeux, J., Schadauer, K., Ståhl, G., Tomter, S., Tomppo, E., Tosi, V., Weiss, P., & Winter, S. (2009). Harmonization of national forest inventories in Europe. Reference definitions. Manuscript.

Tomppo, E., Gschwantner, T., Lawrence, M., & McRoberts, R. E. (Ed.). (2010). National forest inventories: pathways for harmonised reporting. Heidelberg, Springer.

Vidal, C., Lanz, A., Tomppo, E., Schadauer, K., Gschwantner, T., Di Cosmo, L., & Robert, N. (2008). Establishing forest inventory reference definitions for forest and growing stock: a study towards common reporting. *Silva Fennica, 42*, 247–266.

Winter, S., Chirici, G., McRoberts, R. E., Hauk, E., & Tomppo, E. (2008). Possibilities for harmonizing national forest inventory data for use in forest biodiversity assessments. *Forestry, 81*, 33–44.

Chapter 5
Harmonization Tests

Gherardo Chirici, Ronald E. McRoberts, Susanne Winter, Anna Barbati, Urs-Beat Brändli, Meinrad Abegg, Jana Beranova, Jacques Rondeux, Roberta Bertini, Iciar Alberdi Asensio, and Sonia Condés

Abstract Chapter 5 reports the results of testing the proposed procedures for harmonizing estimates of indicators for six of the seven essential features of forest biodiversity. Twenty indicators were tested using data from the common database. In general, positive results were obtained for forest categories, forest structure, forest age, deadwood, and naturalness; the results were less positive for ground vegetation because of the considerable differences in definitions and data acquisition methods. Of importance is, that the test focused on assessing harmonization procedures rather than on producing comprehensive estimates for particular countries or forest categories.

G. Chirici (✉)
Università degli Studi del Molise, Italy
e-mail: gherardo.chirici@unimol.it

R.E. McRoberts
Forest Inventory and Analysis, Northern Research Station, USDA Forest Service, USA
e-mail: rmcroberts@fs.fed.us

S. Winter
Department für Ökologie und Studienfakultät für Forstwissenschaft und Ressourcenmanagement, Technische Universität München, Germany
e-mail: winter@wzw.tum.de

A. Barbati
Università degli Studi della Tuscia, Italy
e-mail: barbati.sisfor@unitus.it

U.-B. Brändli and M. Abegg
Swiss Federal Institute for Forest, Snow and Landscape Reserach (WSL), Schweizerisches Landesforstinventar, Switzerland
e-mail: meinrad.abegg@wsl.ch; urs-beat.braendli@wsl.ch

J. Beranova
IFER, Institute of Forest Ecosystem Research Ltd, Czech Republic
e-mail: iana.beranova@ifer.cz

J. Rondeux
Gembloux Agro-Bio Tech, Université de Liège, Belgium
e-mail: jrondeux@ulg.ac.be

G. Chirici et al. (eds.), *National Forest Inventories: Contributions to Forest Biodiversity Assessments*, Managing Forest Ecosystems 20,
DOI 10.1007/978-94-007-0482-4_5,

5.1 Introduction

The results of the harmonization tests conducted using the raw NFI data from the common database described in Chap. 4 are presented in this chapter. The tests aim to propose and verify possible solutions for harmonizing estimates of indicators for six of the seven essential features of forest biodiversity (Table 2.6). The tests focus on identifying indicators that can be estimated using raw NFI data and developing techniques for harmonizing the estimates across countries. The results illustrate the feasibility of calculating the estimates and, optionally, of assessing the impacts of different harmonization alternatives. Based on the results, the authors also suggest possible improvements in field data collection methods that could increase the number of indicators that could be estimated using NFI data or that could contribute to development of more accurate and simple bridges for use as harmonization techniques.

Selection of the biodiversity indicators and of the harmonization techniques for the tests is limited by the availability of raw NFI data in the common database. The availability of broader thematic and geographic sets of NFI data could lead to different indicators and/or different harmonization methods. In this sense, the proposed tests are data driven. Nevertheless, the final aim of the tests at this stage is not to identify the best approaches to harmonization but rather to present feasible approaches through development of bridges that span the differences between local definitions and international reference definitions.

Because the database used for the tests does not consist of representative samples from individual countries or specific forest categories, the results cannot be used to infer values of biodiversity indicators for countries or forest categories but must be constrained to the domain of the data.

Harmonization of estimates for all indicators is based on a per plot harmonization approach. With this approach, estimates of biodiversity indicators are calculated for each NFI plot represented in the common database, and the plot-level estimates are then generally aggregated by country or by forest category. One effect of this approach is that many of the indicators estimated in the test may suffer from a lack of comparability due to different NFI plot sizes and different measurement thresholds. To compensate, tests for some indicators are restricted to plots of similar size and to observations for variables with common measurement thresholds. In the Conclusion (Sect. 5.8) possible solutions are reported.

R. Bertini
Università degli Studi di Firenze, Italy
e-mail: bertini.roberta@gmail.com

I.A. Asensio
Departamento Sistemas y Recursos Forestales, Centro de Investigación Forestal, Instituto Nacional de Investigación y Tecnología Agraria y Alimentaria, Spain
e-mail: alberdi.iciar@inia.es

S. Condés
Escuela Técnica Superior de Ingenieros de Montes, Universidad Politécnica de Madrid, Spain
e-mail: sonia.condes@upm.es

In Sect. 5.2, Barbati classifies European NFI plots available in the common database according to the European Forest Types (EFT) system of nomenclature developed by the European Environmental Agency (EEA 2006). The EFT system can be considered an indicator for reporting area by forest category, but it is also frequently used as a source of strata for which aggregated estimates of the other indicators are reported.

In Sect. 5.3, Winter and McRoberts present indicators for estimation of forest vertical, horizontal and compositional diversity. The indicators are based on simple tree-level data and demonstrate that observations and measurements of traditional forest variables commonly acquired by NFIs may be used for harmonized estimation of biodiversity indicators using relatively simple bridges.

In Sect. 5.4, Brandli, Abegg, and Beranova illustrate indicators for three variables of potential use for evaluating the age of a forest stand using plot-level data. The authors investigated different methods for estimating the different indicators and the effects of different local field methods. Because tree age is frequently not assessed by NFIs, the possibility of substituting diameter at-breast-height (dbh) for age was also considered.

In Sect. 5.5, Rondeux, Bertini, and Chirici present the results of a deadwood test to assess the effects of three different reference definitions distinguished by different minimum dbh thresholds on estimates of total deadwood volume per unit area and estimates by classes of vertical and horizontal spatial position, species, and decay stage.

In Sect. 5.6, Alberdi Asensio, Condés, Bertini, and Winter propose indicators for ground vegetation. Because information related to ground vegetation collected by NFIs is frequently less detailed and less standardized than those for tree components, harmonization is particularly difficult and complex. Therefore, three relatively simple indicators based on ground vegetation, shrub coverage and shrub presence/absence are proposed.

In Sect. 5.7, McRoberts, Chirici, and Winter propose a new combined index for forest naturalness based on three components: species richness, estimated using the Shannon index; horizontal structural diversity, using the standard deviation of tree dbh; and harmonized estimates of deadwood volume per unit area.

The indicators adopted for the tests reported in Chap. 5 are summarized in Table 5.1.

5.2 Forest Categories

5.2.1 Introduction

European forests are highly variable with respect to ecological conditions and levels of anthropogenic modification. Given this variability, it is very difficult to grasp the meaning of data on biodiversity variables and their trends apart from an appropriate ecological context. Larsson et al. (2001) assert that variation in biodiversity conditions across European forests (Sect. 3.2) may be attributed to a combination of *key factors*: structural characteristics such as tree species and age, old growth forest left for free development, compositional features such as presence of native versus

Table 5.1 Indicators tested

Essential feature	Indicator
Forest categories	1.1 Forest category according to the system of nomenclature developed by the EEA (2006)
Forest structure	2.1 Relative abundance of native tree species in terms of basal area
	2.2 Number of native tree species
	2.3 Proportion of plots with 1, 2, 3 and more native tree species
	2.4 Largest diameter trees
	2.5 Standard deviation of the tree heights
	2.6 Number of vertical layers
	2.7 Frequency distribution of standard deviation classes of dbh
	2.8 Shannon index for tree species
Forest age	3.1 Dominant age
	3.2 Mean age
	3.3 Weighted mean age
	3.4 Old trees
Deadwood	4.1 Deadwood volume by decay class, tree species, and horizontal/vertical position
Ground vegetation	5.1 Cover of ground vegetation
	5.2 Cover of shrub species
	5.3 Presence/absence of shrub species
	5.4 Presence/absence of shrub genus
	5.5 Presence/absence of ground vegetation life forms
Naturalness	6.1 Naturalness combined index

non-native tree species and associated plant and animal species composition, and functional factors such as abiotic/biotic disturbances and forest management.

The combinations of key factors vary in different regions and sites of Europe due to natural biogeographic, bioclimatic, and ecological factors and to anthropogenic factors such as local modification of forest composition including introduction of non-native species. During the last decade, this variation in biodiversity conditions has motivated European experts to develop a pan-European forest typology based on the interplay of the key factors affecting biodiversity that can be used to stratify European forests into a limited number of homogeneous forest types (Barbati and Marchetti 2004; Bradshaw and Møller 2004). To achieve this goal, a scheme of European Forest Types (EFT) has been proposed with the objective of improving future reporting of the state of European forests for processes such as the MCPFE (EEA 2006; Barbati et al. 2007).

The EFTs were specifically conceived to facilitate improved understanding, interpretation and communication of data on forest conditions, including biodiversity, by enabling comparisons of ecologically similar forests. Other available forest classification schemes such as the Eunis Habitat Classification (Davies et al. 2004) and the phytosociological alliances (Rodwell et al. 2002), are not suitable for this purpose. In fact, both these classifications have impractically large numbers of classes: 52 Eunis level III classes and 110 alliances. In addition phytosociological alliances do not include forest formations of purely anthropogenic origin such as plantations.

The EFTs comprise 14 main *forest categories* (Table 5.2) reflecting variation in the main factors that affect European forest biodiversity variables such as forest

Table 5.2 European forest categories

Category	Name	Main characteristics
1	Boreal forest	Extensive boreal, species-poor forests, dominated by *Picea abies* and *Pinus sylvestris*. Deciduous trees including birches (*Betula* spp.), aspen (*Populus tremula*), rowan (*Sorbus aucuparia*) and willows (*Salix* spp.) tend to occur as early colonisers
2	Hemiboreal and nemoral coniferous and mixed broadleaved-coniferous forest	Latitudinal mixed forests located in between the boreal and nemoral (or temperate) forest zones with similar characteristics to cat. 1, but a slightly higher tree species diversity, including also temperate deciduous trees like *Tilia cordata*, *Fraxinus excelsior*, *Ulmus glabra* and *Quercus robur*. Includes also: pure and mixed forests of natural origin dominated by *Pinus sylvestris* (Scotland and central Europe) and the anthropogenic coniferous forests originated from plantations of *Picea abies* or pines of the *Pinus nigra* group in the nemoral forest zone
3	Alpine coniferous forest	High-altitude forest belts of central and southern European mountain ranges, covered by *Picea abies*, *Abies alba*, *Pinus sylvestris*, *Pinus nigra*, *Larix decidua*, *Pinus cembra* and *Pinus mugo*
4	Acidophilous oak and oak-birch forest	Scattered occurrence associated to less fertile soils of the nemoral forest zone; the tree species composition is poor and dominated by acidophilous oaks (*Q. robur*, *Q. petraea*) and birch (*Betula pendula*)
5	Mesophytic deciduous forest	Related to medium rich soils of the nemoral forest zone; forest composition is mixed and made up of a relatively large number of broadleaved deciduous trees: *Carpinus betulus*, *Quercus petraea*, *Quercus robur*, *Fraxinus* spp., *Acer* spp. and *Tilia cordata*
6	Beech forest	Widely distributed lowland to submountainous beech forest. Beech, *Fagus sylvatica* and *F. orientalis* (Balkan) dominate
7	Mountainous beech forest	Mixed broadleaved deciduous and coniferous vegetation belt in the main European mountain ranges. Species composition differs from cat. 6, including spruce, fir, birch and further mesophytic deciduous tree species
8	Thermophilous deciduous forest	Deciduous and semi-deciduous forests mainly of the Mediterranean region dominated by thermophilous species. Mainly of *Quercus* spp.; *Acer* spp., *Ostrya* spp., *Fraxinus* spp., *Carpinus* spp. are frequent as associated secondary trees
9	Broadleaved evergreen forest	Broadleaved evergreen forests of the Mediterranean and Macaronesian regions dominated by sclerophyllous or lauriphyllous trees, mainly *Quercus* species
10	Coniferous forests of the Mediterranean, Anatolian and Macaronesian regions	Varied group of coniferous forests in Mediterranean, Anatolian and Macaronesian regions, from the coast to high mountains. Dry and often poorly-developed soils limit tree growth. Several tree species, including a number of endemics, of *Pinus*, *Abies* and *Juniperus* species

(continued)

Table 5.2 (continued)

Category	Name	Main characteristics
11	Mire and swamp forest	Wetland forests on peaty soils widely distributed in the boreal region. Water and nutrient regime determines the dominant tree species: *Pinus sylvestris*, *Picea abies* or *Alnus glutinosa*
12	Floodplain forest	Riparian and riverine species-rich forests characterized by different assemblages of species of *Alnus*, *Betula*, *Populus*, *Salix*, *Fraxinus*, *Ulmus*
13	Non-riverine alder, birch or aspen forest	Pioneer forests dominated by *Alnus*, *Betula* or *Populus* species
14	Forest of exotic tree species	Plantations and self-sown forest dominated by introduced tree species non-native to Europe (*Eucalyptus spp.*, *Robinia pseudoacacia*, *Acacia spp.*, *Picea sitkensis*, *Pinus contorta*, *Pseudotsuga menziesii*, *Tsuga heterophylla*). Occur on a wide range of site conditions which otherwise would develop forests of above categories

structure, tree and plant species composition, and deadwood levels. Notably, forest categories 1–10 and 13 correspond to groups of ecologically distinct forest communities of varying breadths that are dominated by specific assemblages of trees native to Europe. The forest physiognomies of forest categories 1–10 are mainly determined by the latitudinal/altitudinal zonation of European vegetation and by climatic and edaphic variation. Forest categories 11–12 and 13 include azonal forest communities, and forest category 14 includes forest stands consisting primarily of exotic, non-native, or introduced tree species established as forest plantations or originating from natural regeneration.

The concept of forest categories has already been used to interpret biodiversity data obtained from forest monitoring plots distributed throughout Europe (Fischer et al. 2009). In addition, a simplified version of the scheme has been used to analyze climate change impacts separately for different bioclimatic regions and forest types of Europe (Lindner et al. 2009).

Within COST Action E43, the 14 forest categories have been used as a reference scheme for classifying NFI ground plots based on data submitted by 13 European countries. The primary goal of the harmonization test is to evaluate the potential of the European forest categories as a reference scheme for classifying large sets of European NFI plots using existing NFI data. Of particular importance, the harmonization test goal is not to provide reliable statistics on the relative frequency of forest categories either at country or European levels. The rationale behind this post-stratification of NFI plots is to facilitate analyses of harmonized estimates of biodiversity indicators in a meaningful way, i.e., with respect to ecologically homogeneous strata of European relevance which is how forest categories are defined.

5.2.2 Methods

The 13 countries (Table 5.3) contributing data for the harmonization test on forest categories collectively represent all European biogeographical regions. Each country selected a subset of their full sample of NFI ground plots for inclusion in the common database (Chap. 4). The sizes of the subsets ranged from 300 to 1,460 plots per country, for a total of 8,423 plots. The harmonization test consisted of classifying the NFI plots represented in the common database using the classification scheme represented by the 14 European forest categories as an external reference. The classification scheme, including key classification factors and detailed descriptions of categories, was reported by the European Environment Agency (EEA 2006). The objective, as indicated in Sect. 3.2, is to bypass the difficulties of harmonization of forest categories using the various schemes currently adopted within European NFIs. The decision criteria for assigning NFIs plots to forest categories are based on variables commonly assessed by NFIs or variables whose values can be obtained from other external sources:

- tree species native to Europe as identified in the Atlas Florae Europeae database (http://www.fmnh.helsinki.fi/english/botany/afe/index.htm);
- dominant tree species defined as the single or group of tree species accounting for more than 50% of basal area;
- locations of the plots with respect to the biogeographical regions of Europe (http://dataservice.eea.europa.eu/atlas/viewdata/viewpub.asp?id=2038); this information is required by the classification rules to distinguish the main coniferous-dominated European forest categories (Forest Categories 1, 2, 3, 10);
- locations of plots with respect to potential natural vegetation delineated in the European map of natural vegetation (Bohn et al. 2000); this information is used primarily for assessing forest site ecological characteristics used in the key classification factors (bioclimatic and forest vegetation zones/altitudinal belts, soil fertility and hydrological regimes); although some inconsistency between the physiognomies of actual and potential forest vegetation is acknowledged in the definitions of forest categories (e.g. Forest Category 2, nemoral coniferous forests), reference to the natural forest vegetation zones was still useful for purposes of guiding the classification process;
- proximity to major European rivers, particularly for Forest Category 12, floodplain forest.

The NFI plots were classified by each country using the classification rules set out by the European forest categories nomenclature which also served as a basis for country-level decision flow-charts. The flow charts were necessary to establish bridges from country-level information on the variability in forest composition and species mixtures, both assessed using basal area data, to classes of ecological relevance at the European level. As an example, the flowchart developed by the Irish NFI is shown in Fig. 5.1.

Table 5.3 Classification of the subsets of NFIs ground plot by forest category (EEA 2006)

	Country												
Category	Austria	Belgium	Czech Republic	Denmark	Finland	Germany	Ireland	Italy	Norway	Portugal	Spain	Sweden	Switzerland
1					X				X			X	
2	X		X			X	X		X				
3	X					X		X					X
4		X	X				X						
5	X	X	X	X		X	X	X					
6	X	X	X	X		X	X						X
7	X		X			X		X			X		X
8								X		X	X		
9								X		X	X		
10								X		X	X		
11			X		X	X	X						
12	X			X		X	X	X					
13			X		X	X	X		X			X	
14	X	X	X	X	X	X	X	X		X	X	X	
Tot[a]	7	4	8	4	4	9	8	8	3	4	5	3	3

[a] Total number of categories in the data available in the common DB (Chap. 4)

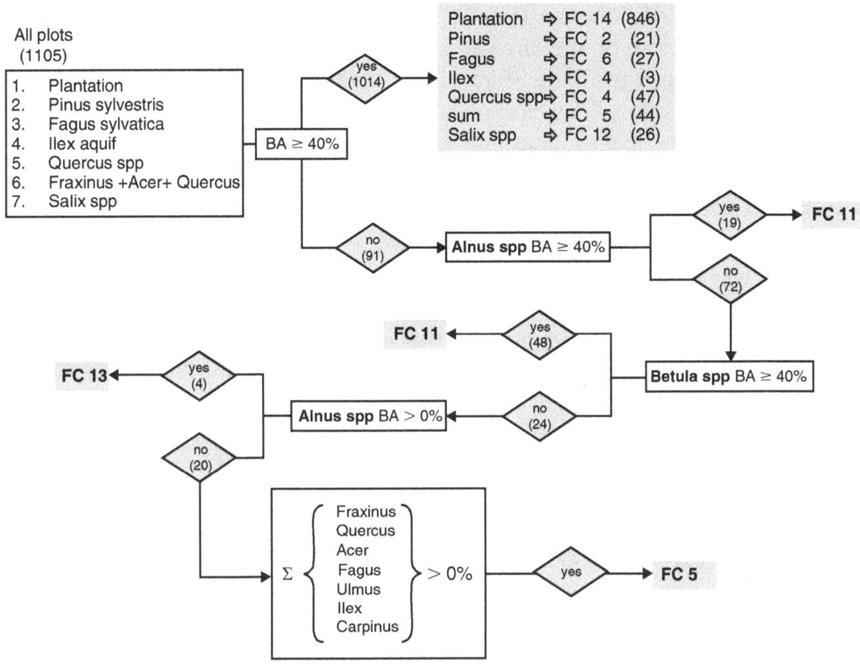

Fig. 5.1 Decision flow-chart used to classify Irish NFI plots according to the European forest categories (EEA 2006)

5.2.3 Results and Discussion

All countries were able to classify almost all their subsets of NFI plots with respect to the European forest categories (Table 5.3) with support from the decision flow charts. The results of the harmonization test are promising. In particular, they show that use of the European forest categories as a reference is feasible without collecting additional field data, at least for the countries in the test. The lessons learned from this exercise provide guidance for developing harmonized forest classification schemes in other regions of the world that are characterized by large scale variability in forest conditions due to ecological and human factors.

5.3 Forest Structure

5.3.1 Introduction

The insights from the analyses of forest structure variables (Sect. 3.3) are that the individual tree measures such as dbh, height and species are the most commonly assessed variables by national forest inventories (NFI) in Europe and the United States of America (USA). Thus, they are also the most promising variables for reporting on the large-scale structural aspects of forest biodiversity. The primary rationale for basing forest biodiversity indicators on NFI data is that the huge existing national databases can be readily used for estimation purposes.

This basic approach to assessing forest biodiversity does not require new monitoring programmes or even new monitoring concepts. Rather, acquisition of a sufficient database of information for forest biodiversity assessments should require only minor modifications to existing programmes such as addition of new variables or revision of measurement procedures. If successful, this approach would be cost efficient with respect to establishing a large-scale and European-wide forest biodiversity monitoring programme. This basic approach can be extended by adding monitoring modules that focus on features such as spatial landscape composition, avifauna, species lists of the ground vegetation (Sect. 5.6), or other species that may be not represented by general structural indicators. However, rare species such as the northwest population of the Siberian crane *Grus leucogeranus* in Russia (BirdLife International 2000), for which habitat requirements and specific reasons for their decline are not well known, will likely become extinct without specific intervention. Such species require additional and specific monitoring attention. In general, population trends for rare species cannot be satisfactorily monitored using structural indicators (Pearman et al. 2006).

The suitability of common forest structure variables for assessing the biodiversity status of forests on a general level with simple indicators is investigated and reported in this section by forest categories (EEA 2006). The underlying assumption is that the forest categories 1–10 (Table 5.2) describe the zonal forest tree species distribution and communities and represent a broad climatic gradient from the colder and wetter northeast to the warmer and drier southwestern climate zones. Additionally, forest categories 11–13 that originate in azonal abiotic habitat conditions such as floodplain forests and mire and swamp forests are considered separately. Finally, inventory data were submitted from the USA for five forest types (U.S. Forest Service 2007) from three ecoregions (Bailey 1976, 1983) (Fig. 4.1): Loblolly pine (*Pinus taeda*, forest type 161) from parts of ecoregion 231 in the southeastern plantation region, Douglas fir (*Pseudotsuga menziesii*, forest type 201) from ecoregion 242 in the northwestern rainforest region, and Aspen (*Populus tremuloides* and *P. grandidentata*, forest type 901), Paper birch (*Betula papyrifera*, forest type 902), and Balsam poplar (*Populus balsamifera*, Forest Type 904) from parts of ecoregion 212 in the north central Lakes States region.

Harmonization of estimates of forest biodiversity indicators uses three types of bridges constructed to accommodate differences between national and international reference definitions (Vidal et al. submitted, Ståhl et al. submitted). Reductive and neutral bridges may be used without changing country NFI methods, but expansive bridges may include models and/or acquisition of additional data. Our harmonization tests feature reductive and neutral bridges; where expansive bridges are not feasible, construction of a new reference definition or standardization of definitions and measurement protocols is recommended.

5.3.2 Methods

Nativeness of tree species: Tree species are defined as native if they are included in the EEA (2006) description of Forest Categories 1–13 (Table 5.2). For tree species represented in the common database but not included among the EEA descriptions, we acquired additional information from the literature on their natural distributions by forest categories. Any assertions that tree species were not native in a particular forest category were carefully checked. The EEA (2006) decision trees require that basal areas of native species exceed 50% for determining forest categories. Because dbh was easy to harmonize (Sect. 5.3.3), basal area could be estimated for the test.

Diameter at breast height (dbh): The largest minimum dbh threshold for countries that submitted data for the common database was 120 mm for both Cyprus and Switzerland (Fig. 5.2). We used this largest minimum dbh threshold as a reductive bridge to harmonize the dbh country data for the harmonization tests.

Layer harmonization: National approaches to assessing height layers were characterized with respect to the number of layers used: one layer, two layers, and three or more layers which was also characterized as multi-layering (Table 3.2). For the harmonization tests, national approaches to assessing layers were assumed to follow the natural heterogeneity of forest structures that develop in different forest categories.

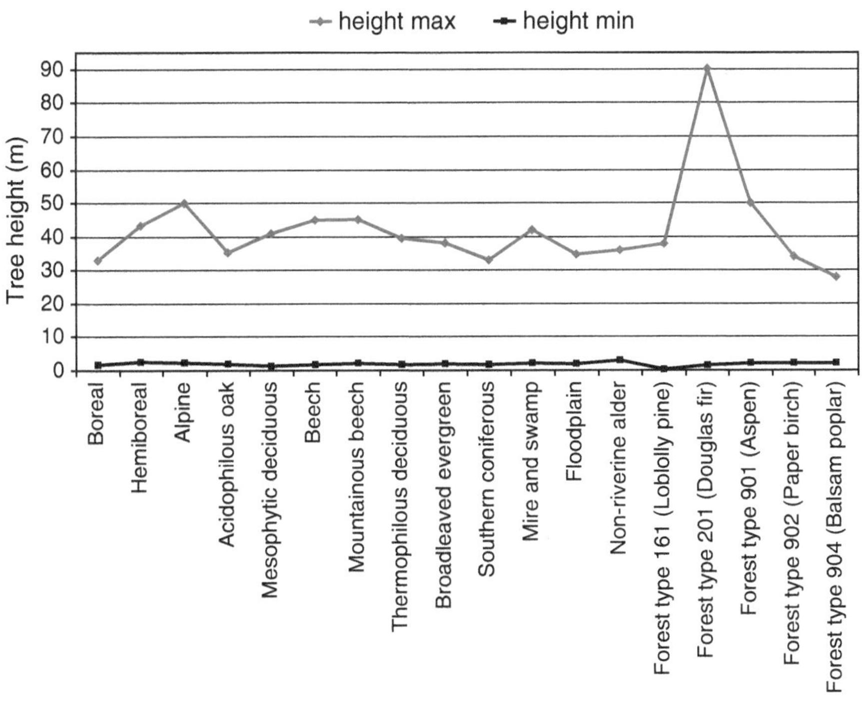

Fig. 5.2 Minimum and maximum heights for European forest categories (Table 5.2, EEA 2006) and American forest types (U.S. Forest Service 2007)

Construction of a reference definition for layers is based on the assumption that national definitions were constructed using different approaches. For example, in Finland the layer definition specifies that the number of recruits must be silviculturally sufficient to support establishment of the next mature stand. More than 5,000 recruits per hectare are regarded as sufficient for a regeneration layer. For other countries, minimum coverage for layers can be less than 5,000 recruits. We followed two steps for construction of a reference definition for layer. First, we reviewed minimum and maximum tree heights in the common database by forest categories (Fig. 5.2). Increasing maximum tree heights seem not to be correlated with increasingly favourable growth conditions characteristic of some forest categories. Several particular findings support this finding: (1) Alpine forest (Forest Category 3), which includes the greatest elevations of European mountain ranges, has mostly less favourable height growth conditions but a very large maximum height (50 m); and (2) Mire and swamp (forest category 11), also known for limited height growth potential caused by water surpluses and O_2 and nutrition shortages, also had a large maximum height. As a result, we assume that maximum height is mainly influenced by the impacts of management. Forests in both categories, Alpine forest (forest category 3) and Mires and swamps forest (forest category 11), are often less intensively managed and seldom harvested in many countries because of steep slopes, intensive snow cover in alpine regions, and wet and inaccessible lands in mire and swamps. As a result, trees in these categories of forest tend to grow to greater ages and heights than in forest categories with more favourable site and harvesting conditions.

Second, three distinct forest layers were assumed to be appropriate for most forests, although Douglas fir forests with maximum heights greater than 90 m may develop more than three distinguishable layers. Three approaches to constructing a layer reference definition were considered.

- Approach 1. The minimum tree height (ht) for assignment to a layer was 2 m or 5 m; the three layers were defined as $2\ \text{m} \leq \text{ht} < 10\ \text{m}$, $10\ \text{m} \leq \text{ht} < 20\ \text{m}$, and $\text{ht} \geq 20\ \text{m}$ (alternatively for the first class $5\ \text{m} \leq \text{ht} < 10\ \text{m}$).
- Approach 2. The minimum height was 2 m; the three layers were of equal length and were defined as $2\ \text{m} \leq \text{ht} < 18\ \text{m}$, $18\ \text{m} \leq \text{ht} < 34\ \text{m}$, and $34\ \text{m} \leq \text{ht} < 50\ \text{m}$ where 50 m is the maximum height of the European forests in the common database. This test is derived from the European growth conditions; the exceptionally maximum height of Douglas fir in the ecoregion USA 2 of 90.2 m was not considered to calculate the equal length of the layers.
- Approach 3. The minimum tree height was 2 m; the three layers were of equal length where the length was one-third the maximum tree height observed within each forest category.

The first two approaches are based on the assumption that differences in maximum tree heights within forest categories are influenced primarily by forest management rather than by natural growth conditions. The third approach is based on the assumption that different growth conditions within forest categories could be also the primary factor influencing maximum tree heights.

5.3.3 *Results and Individual Indicator Discussion*

On average, approximately three-fourths (74%) of the plots in the common database were used for the different harmonization and indicator tests (Table 5.4). After use of a reductive bridge based on a minimum dbh threshold of 120 mm and a minimum height of 2 m, 87% of the plots remained. Because several countries did not contribute layer data, data for only 33% of the plots represented in the common database could be used for the harmonization tests (Table 5.4).

The data for 100% of the Swiss plots represented in the common database could be used for the harmonization tests because the large Swiss minimum dbh of 120 mm was the threshold used for the reductive bridge and because Swiss data were contributed for all the other variables: native tree species, forest layers, and tree heights. Large percentages of plots could also be used for Spain (99%), the Czech Republic (95%) and Germany (93%). Additionally, application of the reductive bridge led, on average, to retention of 70% of the plots represented in the common database. This result suggests that the proposed reductive bridges do not, in general, lead to important loss of NFI information, although further tests should be conducted on complete national datasets. This harmonization of dbh and height, and of the tree species using the ICP codes (Sect. 5.7), enabled testing of eight indicators.

5.3.3.1 Indicator 1: Percentage of Basal Area in Native Tree Species

The indicator specifies the general tree species composition of forests in Europe and the USA. Remarkably, only slightly more than 50% of basal area is for native tree species in most European forest categories; two forest categories (hemiboreal forest, and mire and swamp forest) have slightly less than 50% (Fig. 5.3). Two conclusions may be drawn from this finding: (1) the biodiversity and distribution of native tree species is greatly reduced in the tested parts of European forests, and (2) the guidelines for assigning plots to forest categories were not followed properly. For all forest categories, assignment to categories following the EEA guidelines (EEA 2006) were to be based on the dominant species, i.e., the species that accounted for more than 50% of the basal area. The second conclusion suggests that the assignment of plots to forest categories may have been based more on the number of trees per species than on basal area per species.

The overall conclusion is that this indicator requires only knowledge of native tree species by forest categories. Country data provided by species groups are not suitable for estimation of this indicator. Use of the percentage of basal area focuses on determining the dominant tree species. Time series of observations indicate trends in tree species composition and biodiversity for large forest regions. To facilitate estimation of the indicator, European NFIs should report tree species using the ICP plant species codes (Chap. 4, Sect. 5.6).

Table 5.4 Numbers and proportions of plots used for the harmonization tests

Country	Total number of plots in common DB	Percentage of country plot number					
		Native tree species	dbh threshold 120 mm	Height threshold 2 m	Height threshold 5 m	Country layer definition	Mean
Austria	1,104	63	60	63	62	100	70
Belgium	400	37	90	58	58	0	48
Czech Republic	302	87	94	96	96	100	95
Denmark	1,458	32	92	98	92	2	63
Finland	336	88	76	68	62	100	79
Germany	790	89	91	93	93	100	93
Ireland	1,105	70	90	100	96	0	71
Italy	351	100	93	96	89	71	90
Norway	1,361	53	65	53	52	48	54
Portugal	250	64	67	56	56	0	49
Spain	775	100	97	100	96	100	99
Sweden	188	80	80	74	74	100	82
Switzerland	401	100	100	100	100	100	100
USA	5,752	94	93	94	93	0	75
All	14,573	77	87	87	85	33	74

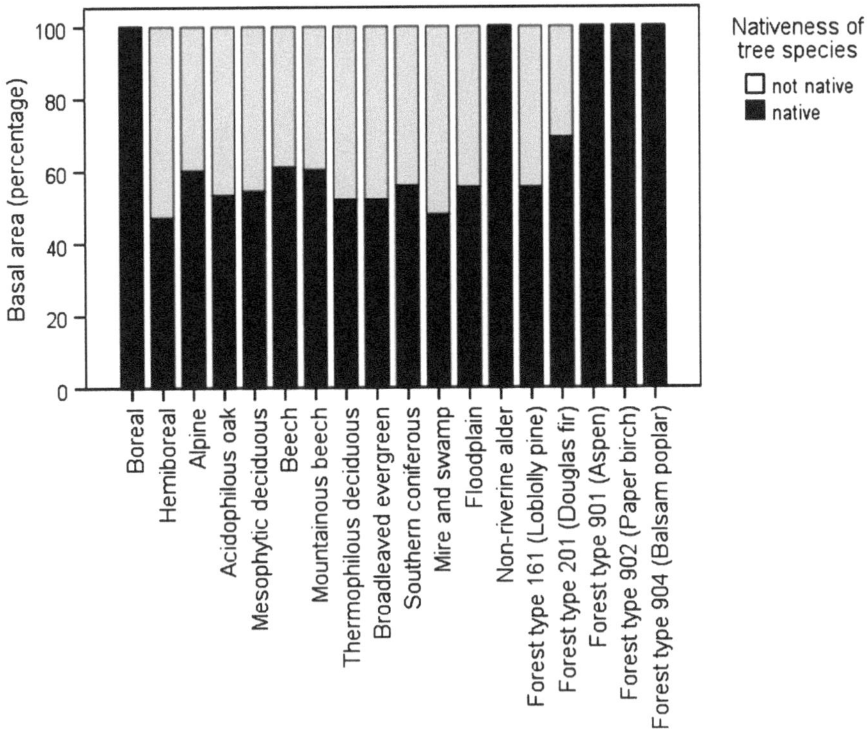

Fig. 5.3 Percentage of basal area in native and non native tree species by European forest categories (Table 5.2, EEA 2006) and American forest types (U.S. Forest Service 2007)

5.3.3.2 Indicator 2: Number of Native Tree Species

The diversity of native tree species expressed by the number of different tree species in the forest categories (Fig. 5.4) is a slight modification of the first indicator.

Although caution is necessary because of the limited number of plots represented in the common database, the number of native tree species might increase from the Boreal forest (forest category 1) in the more northern and colder zone to the Mountainous beech forest (forest category 7) in the warmer temperate zone of central Europe (Fig. 5.4). The southern and warmer forests and the azonal forests all have less tree species diversity than the temperate zones. Plots of the five forest types in the three ecoregions of the USA are rich in tree species with approximately four tree species per plot compared with slightly fewer than two for Europe, with the exception of Mountainous beech (forest category 7). However, tree species diversity is influenced by plot size (McRoberts et al. submitted), a factor that was not considered in these tests. Thus, harmonization efforts would benefit from an ecological-based plot size approach. Plot size would not necessarily be standardized but would be selected with consideration given to the natural density and

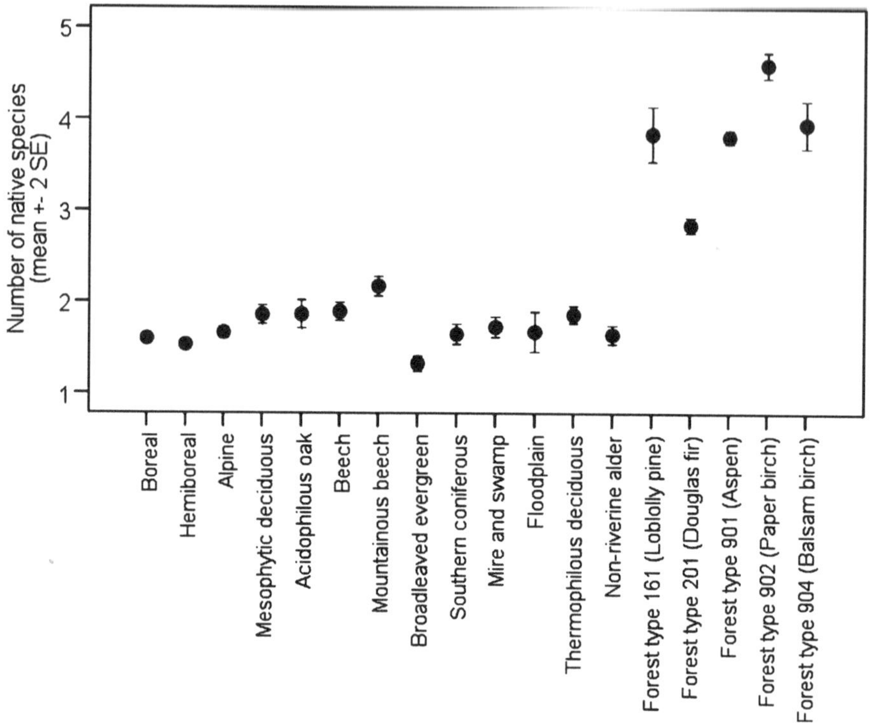

Fig. 5.4 Native tree species per plot (mean and 2-standard error) by European forest categories (Table 5.2, EEA 2006) and American forest types (U.S. Forest Service 2007)

distribution of tree species in the different forest categories. The conclusion is that the diversity of native tree species can be estimated for large areas using the first two indicators (Figs. 5.3 and 5.4).

5.3.3.3 Indicator 3: Proportion of Plots with 1, 2, 3 and More Native Tree Species

This indicator is based on the distribution of plots relative to plot-level tree species diversity (Fig. 5.5) which is simpler and visually more easily understood than indicator 2.

For all European forest categories, more than 60% of the common database plots include two or fewer tree species per plot. In addition, for eight of the 13 forest categories, more than 80% of plots have two or fewer tree species. By contrast, the different ecoregions of USA have fewer than 50% plots with only one or two tree species. The percentages of plots with one or two native tree species decrease from the Boreal forest (forest category 1) in the colder more northern European regions to the Mountainous beech forest (forest category 7) in the temperate zone of central Europe. Broadleaved evergreen forest (forest category 9) in more southern regions

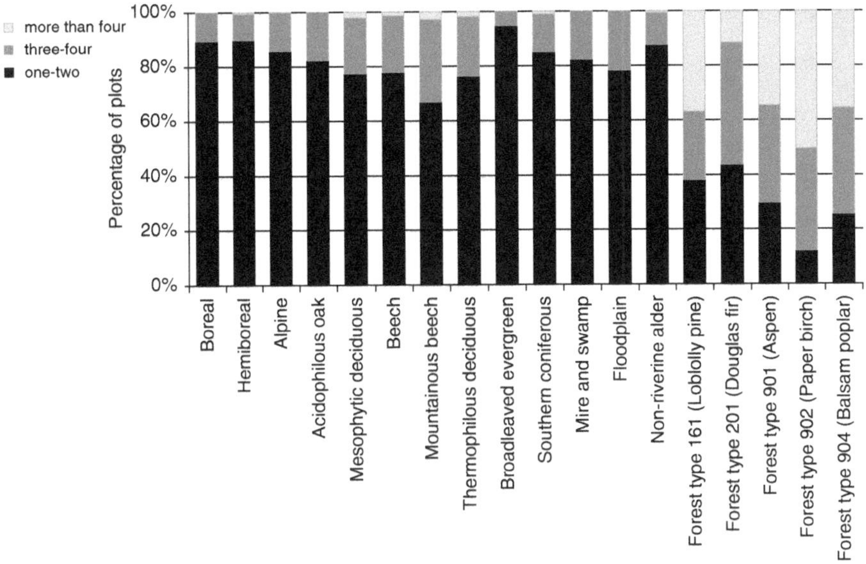

Fig. 5.5 Tree species diversity in plots reported as proportion of plots with different native tree species by European forest categories (Table 5.2, EEA 2006) and American forest types (U.S. Forest Service 2007)

of Europe with the smallest mean number of native species per plot and the greatest proportion of plots with only one or two species is somewhat of an exception to the trends noted in Figs. 5.4 and 5.5. The Mediterranean evergreen oak, a forest type within the Broadleaved evergreen forest category, is naturally dominated by a large number of oak species and represents one of the main potential natural forests of the meso-Mediterranean vegetation region. Today, only small remnants of natural forest communities still exist in Europe. Most forests have been converted to pasture or other open land. Additionally, the remnants with oaks have only a small number of species per plot. A similar result characterizes the Lauriphyllous evergreen forest type of the warm temperate humid zones of Macaroniesia in the Broadleaved evergreen forest (forest category 9). Although the plots represented in the common database are not necessarily representative of the forest categories, the conclusion is that the primary cause of reductions in native species diversity is conversion from natural forest to other uses. This finding for this indicator is similar to that for indicator 2. We tested both indicators to facilitate a choice for large-scale reporting.

5.3.3.4 Indicator 4: Largest Diameter Trees

For this indicator the only harmonization requirement is conversion of measurement units; we used mm for the test. For example, Spain reports dbh in cm which was then converted to mm. For constructing a reference definition for the indicator *Largest*

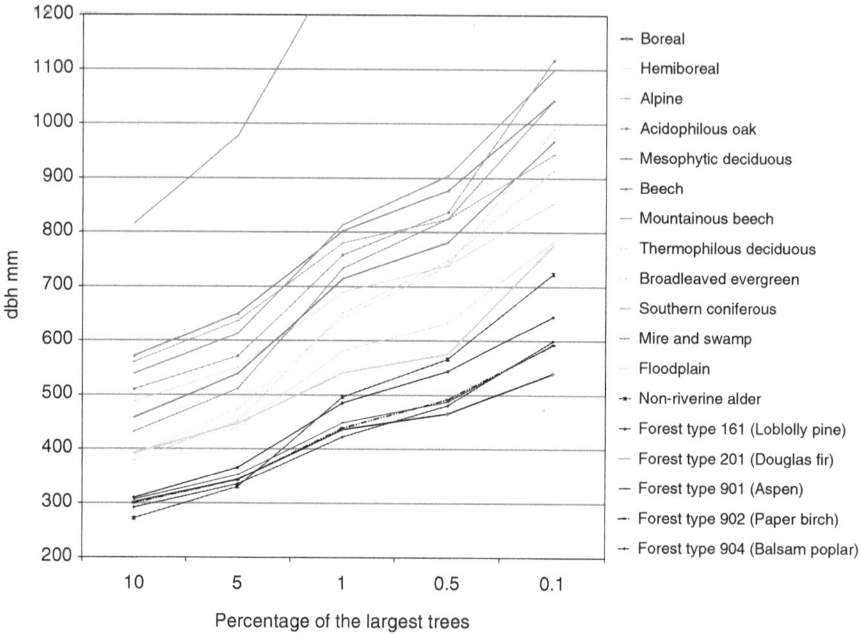

Fig. 5.6 Large dbh (mm) trees by European forest categories (Table 5.2, EEA 2006) and American forest types (U.S. Forest Service 2007)

diameter trees, we estimated mean dbh for different percentile categories of the largest trees: upper 10%, upper 5%, upper 1%, upper 0.5% and upper 0.1%. The lower boundary of the upper 10% of the distribution of trees by dbh is 272 mm (Fig. 5.6). However, this dbh is regarded as too small to represent the large trees of old growth forests that provide high quality habitat and contribute to greater biodiversity. Small trees grow mainly in the boreal forest, non-riverine alder and in four of the five American forest types. The lower boundary of the upper 0.1% of the distribution of trees by dbh is 540 mm which better characterizes old growth forests at large landscape scales and different growth conditions. However, very few trees are included in this small portion of the distribution; for example, there are only 18 such trees for the boreal dataset which includes 2,488 trees. The upper 1% of the distribution has a lower dbh threshold of 423 mm and includes five times more trees. Although the common database includes only a portion of existing NFI data for the boreal regions, the upper 1% of the dbh distribution was selected for development of the indicator.

For Boreal forest (forest category 1) and the American forest types, with the exception of Douglas fir forest (forest type 201) in the northwestern region of the USA, the largest trees are relatively small with dbh less than 500 mm (Fig. 5.7). In the temperate zone, Mountainous beech forest (forest category 7) includes the largest trees of the zonal vegetation and the Mire and swamp forest (forest category 11) of the azonal vegetation. Exceptionally large trees are found in the Douglas fir forest in USA where the mild, wet climate most of the year favours growth in both virgin and managed forests.

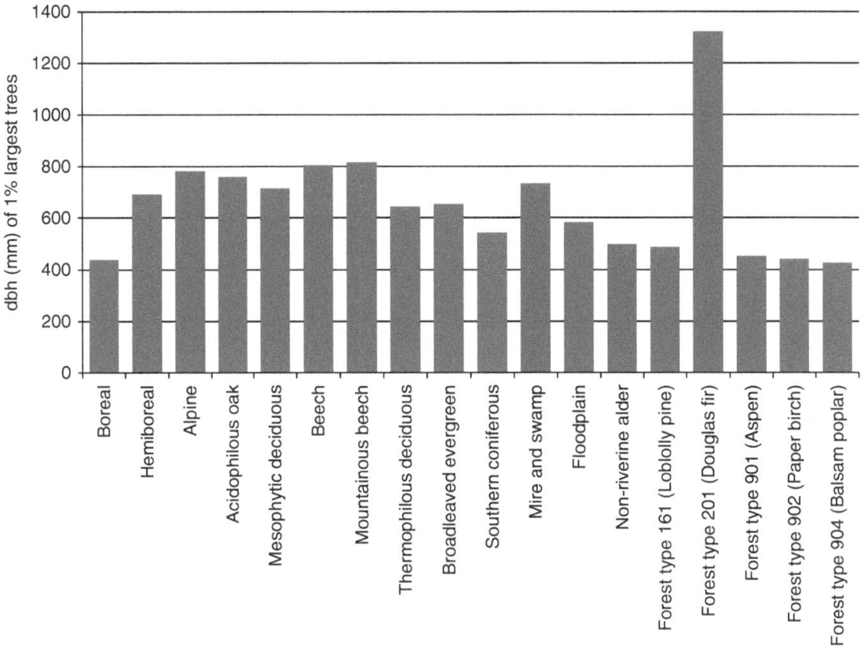

Fig. 5.7 Mean dbh (mm) of the upper 1% of largest trees by European forest categories (Table 5.2, EEA 2006) and American forest types (U.S. Forest Service 2007)

For the data in the common database, the largest 1% of trees is the most suitable for reporting on large trees. However, for largest tree reporting based on the totality of data for European NFIs, the upper 0.1% of the largest trees would likely be more suitable.

5.3.3.5 Indicator 5: Standard Deviation of the Tree Heights

The plot-level standard deviation of tree heights was used as an indicator of forest structural diversity. Two approaches for constructing a reference were tested: (1) all trees with measured or estimated heights were considered, and (2) only trees with heights of at least 5 m were considered.

Forest categories with small mean dbh for the 1% of the largest trees (Fig. 5.7) tend to have smaller standard deviations of tree heights (Fig. 5.8). Differences between mean plot-level standard deviation calculated for all trees measured using the country minimum height threshold and the mean calculated only for trees satisfying the largest reported minimum height threshold of 5 m (the Slovak Republic, Sect. 3.3) are not large (Fig. 5.8). Only 9,641 trees (4% of trees represented in the common database) had heights less than 5 m which meant that these small trees had little effects on results.

For our test, the standard deviation of tree heights mostly increased with increasing numbers of trees (Table 5.5); conversely, tree heights were similar for plots with

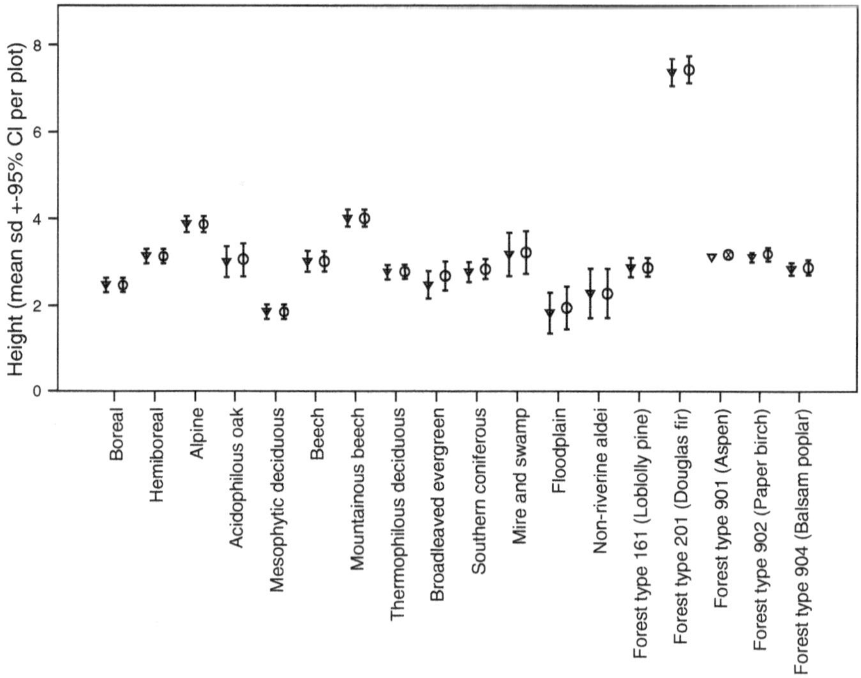

Fig. 5.8 Means of plot-level standard deviations and 95% confidence intervals for plots with at least three trees; triangles: all plot trees in common database were used; circles: only trees with heights ≥ 5 m were used. By European forest categories (Table 5.2, EEA 2006) and American forest types (U.S. Forest Service 2007)

small numbers of trees. These findings were theoretically unexpected. In our test, we found that the greater the number of trees, the greater the maximum tree height; thus, the indicator suggests increasing height diversity which is associated with increasing biodiversity for life-forms such as insects.

5.3.3.6 Indicator 6: Percentages of One-, Two- and Multi-Layered Plots by Forest Category

The percentage of single-layered plots is generally between 40% and 60% for all forest categories (Fig. 5.9). For all forest categories except Beech forests (forest category 6), the percentages of plots that are single layered are at least 40%. Thermophilous forests (forest category 8) have the greatest percentages of single-layered plots with more than 66%. For the ecoregions in the USA, the field protocol for assessing layering is not available. As discussed in Sect. 3.3, the difficulty with harmonization of estimates of layering is that country definitions of layers differ considerably. Thus, the same stand may be assessed as having multiple layers in Germany but only a single layer in Finland. Because the entries in the

Table 5.5 Simple linear regression analyses of the number of trees per plot and the standard deviation of tree heights

European forest category/American forest type	Abbreviated name	Number of plots	Coefficient of the regression	R^2	Significance[a]
Category 1	Boreal	731	0.419	0.232	**0.000**
Category 2	Hemiboreal	689	0.043	0.023	**0.000**
Category 3	Alpine	568	0.110	0.084	**0.000**
Category 4	Acidophilous oak	90	0.027	0.069	**0.012**
Category 5	Mesophytic deciduous	495	0.011	0.004	0.163
Category 6	Beech	390	0.030	0.026	**0.001**
Category 7	Mountainous beech	378	0.018	0.019	**0.007**
Category 8	Thermophilous deciduous	393	0.030	0.162	**0.000**
Category 9	Broadleaved evergreen	165	0.049	0.241	**0.000**
Category 10	Southern coniferous	170	0.019	0.074	**0.000**
Category 11	Mire and swamp	158	0.017	0.200	**0.000**
Category 12	Floodplain	51	0.0	0.000	0.985
Category 13	Non-riverine alder	154	0.232	0.137	**0.000**
Type 161	Loblolly pine	281	–0.009	0.013	**0.059**
Type 201	Douglas fir	1,006	0.006	0.000	0.544
Type 901	Aspen	3,173	0.026	0.063	**0.000**
Type 902	Paper birch	613	0.010	0.017	**0.001**
Type 904	Balsam poplar	246	0.043	0.121	**0.000**

[a] Bold indicates significance at $\alpha = 0.05$

common database were contributed by countries that assessed layers using their own definitions and methods, the layering overview depicted in Fig. 5.9 does not necessarily represent the vertical heterogeneity of European forests. Thus, we tested several methods to construct a layer reference definition based on tree heights (see indicator 7).

5.3.3.7 Indicator 7: Average Number of Tree Layers Per Plot by Forest Category

Potential reference definitions are influenced by the number of trees per plot necessary to constitute a layer (Fig. 5.10a–c). Generally, the average number of layers is approximately 1.5 when a single tree is assumed to be sufficient to form a layer (black lines, Fig. 5.10). However, the average number of layers is clearly less when a minimum of three trees per plot is required to form a layer (grey lines, Fig. 5.10) with the consequence of more plots without any tree layers. Differences in the layer averages are greatest for Boreal forest (forest category 1), the Mire and swamp forest (forest category 11) and Non-riverine alder forest (forest category 13). Stand densities in these latter three forest categories are less than in most other forests because growth is inhibited by climatic and environmental conditions.

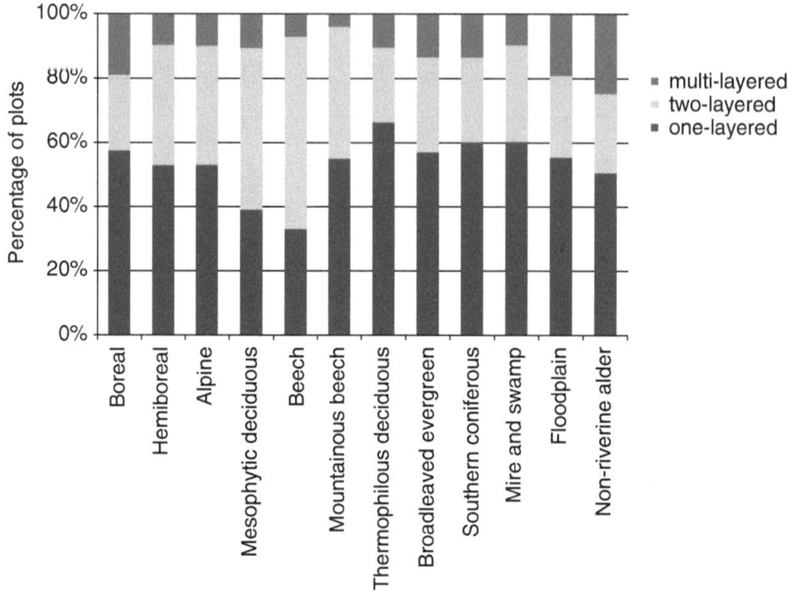

Fig. 5.9 Layering using country definitions; Acidophilous oak (Forest Category 4) omitted because of small numbers of plots with layering information

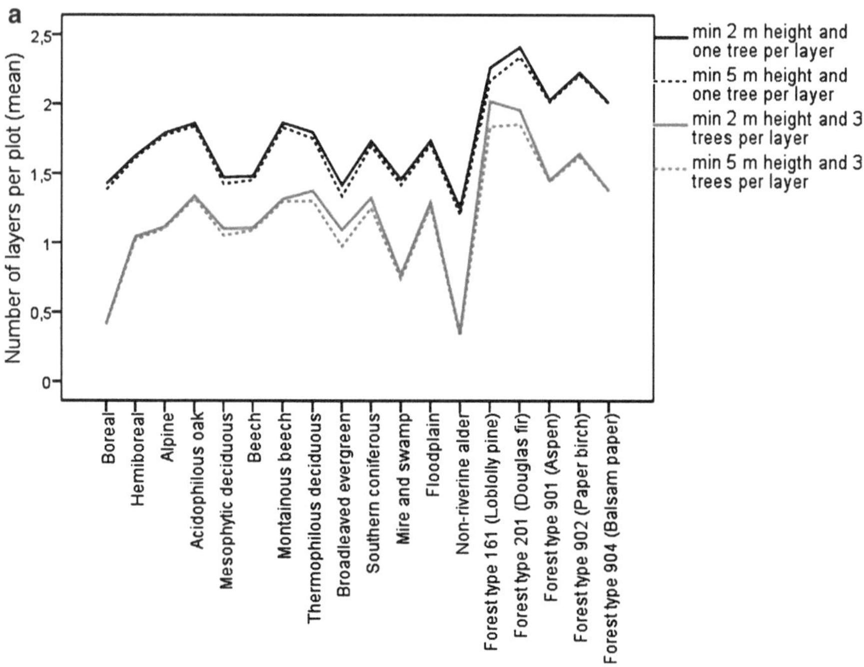

Fig. 5.10 Number of layers per plot: (**a**) top: classes 2<ht≤10 or 5<ht≤10, 10<ht≤20 and ht>20 m; (**b**) middle: equal classes 2<ht≤18, 18<ht≤34, 34<ht≤50 m; (**c**) bottom: three equal interval classes calculated using maximum height for each European forest category (Table 5.2, EEA 2006) and American forest type (U.S. Forest Service 2007)

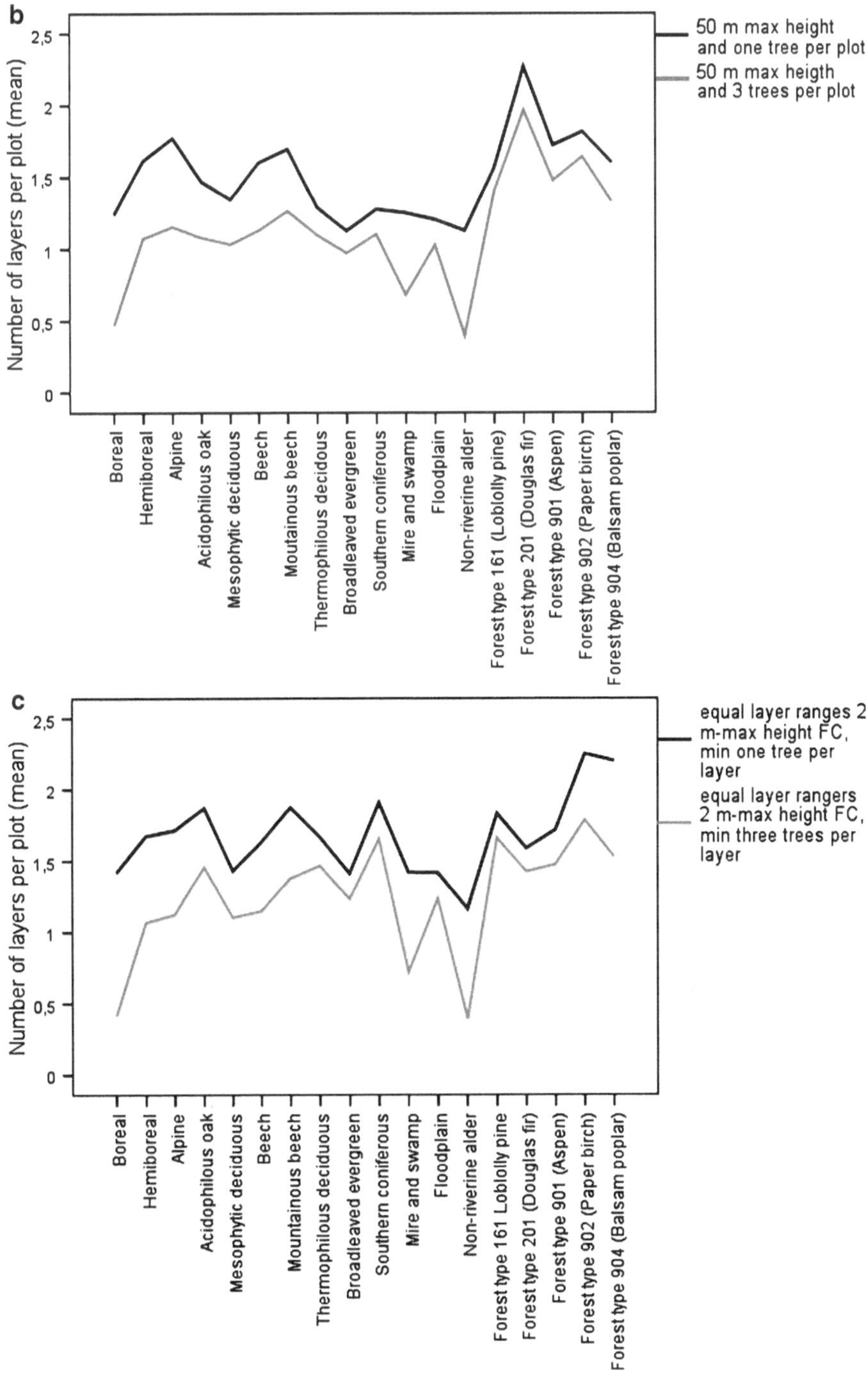

Fig. 5.10 (continued)

The layer height ranges (Sect. 5.3.3) and the minimum height (2 m or 5 m) have less influence on the result than the number of trees per plot (Fig. 5.10). The mean number of layers differs significantly with respect to the 2 m or 5 m minimum height only for the very tall trees in the Douglas fir forest type in the USA (Fig. 5.10a). The number of layers decreases when the layering definition takes into account the maximum height of the specific forest types. Thus, we recommend that the reference definition for layers be based on a division of the maximum height into three equal-length layers.

Greater thresholds for minimum numbers of trees per plot to constitute a layer lead to more plots without any layers. To avoid mean estimates of less than one layer per plot, we propose a requirement of at least one tree per layer for layering assessments. For older trees with larger crown dimensions, acceptance of a single tree per layer seems appropriate, but for young regeneration stands such may not be the case. For purposes of clarity and simplicity, we do not use different layer definitions for different stand densities, although further testing should be considered. The second proposal is to use the layer reference definition based on the maximum heights by forest categories under the assumption that the heights of at least some trees represented in a large database will reach close to the maximum height under natural growth conditions. However, if there are no tall trees for a forest category, selection of the maximum height in the database would be more suitable for purposes of constructing a layer reference definition.

5.3.3.8 Indicator 8: Frequency Distribution of Plot-Level Standard Deviation of dbh

This indicator, which also assesses plot-level or alpha diversity (Whittaker 1972), is addressed in Sect. 5.7 on forest naturalness.

5.3.4 Discussion

Indicators 1–3 on the nativeness of the tree species composition, Indicator 4 on large diameter trees, and Indicator 5 on tree heights are easy to estimate, at least partially because only reductive bridges are necessary. Indicator 6 on layers is the only one that requires more than harmonization among classes. The consequence is that Indicator 6 cannot readily be used for harmonized biodiversity reporting. However, the test using the country layer definitions showed that a general overview of current conditions for European forest categories may be obtained using these country layer definitions. Nevertheless, the assessment of layering is greatly influenced by the definitions; thus, the country definitions should not be used for harmonization purposes. The proposed layer reference definitions using tree heights can easily replace the country layer definitions.

Further data acquisition and indicator testing should focus on the sensitivity of the proposed indicators to different numbers of plots and slight ecological changes such as climate changes or silviculture practices in short time series. Time series of data for testing trends in estimates of the indictors were not available in the common database. However, the 319,728 trees and 14,573 plots represented in the common database were sufficient to test different methods for assessing forest structure and to estimate the variability in forest structure by European forest categories (EEA 2006) and five forest types in the USA (U.S. Forest Service 2007).

An assumption underlying all the tests of biodiversity indicators is that large estimates for the indicators correspond to greater biodiversity. This assumption is justified on the basis of intensively managed forests in Europe and the USA. To evaluate the meaning of changes in estimates of indicators, reference values for forests known to have greater natural structure are necessary for comparison purposes (Sect. 5.7). This approach would permit assessment of the biodiversity relative to nearly natural forest conditions (Winter et al. 2010).

5.4 Forest Age

5.4.1 Categories and Definitions

The ages of trees and stands as well as the age structures of forests are typical and major components of forest ecosystems. In addition, because age can be an important biodiversity indicator as shown in Chap. 2 and Sect. 3.4, forest age was chosen to be one of the essential features of forest biodiversity. For biodiversity purposes, forest age focuses on old forest, is based on the ages of individual trees, and is characterized for this section using three variables: (1) dominant age (age_{dom}), (2) stand age (age_{stand}) and (3) tree age (age_{tree}).

Age_{dom} is defined as the mean age of the 100 trees per hectare with the largest diameters independent of stand or forest structure, tree age distribution or management. Age_{dom} is developed and introduced in a manner analogous to that of the well-known silvicultural variables dominant diameter at-breast-height (dbh_{dom}) and dominant height (H_{dom}) (Pardé 1956; Kändler and Riemer 2005). Because age_{dom} focuses on the largest and oldest trees, it is better related to biodiversity than age_{stand} and is generally independent of diameter measurement thresholds and plot sizes.

Age_{stand} is defined as the mean age of the dominant trees in the stand where dominant trees are in the upper layer of the canopy (Helms 1998). In the case of plantations, stand age is generally based on the year the plantation was established without regard to the age of the nursery stock.

Age_{tree} is defined as the time elapsed since the germination of the seed, the budding of the sprout or the cutting from which the tree developed. In the case of planted trees, the date of planting is the basis for tree age.

5.4.2 Proposed Indicators

Although 90% of European countries assess forest age, existing data are very heterogeneous and permit analyses of age_{stand} at the international level only for even-aged, high forests (Sect. 3.4). For this study, *high forest* is defined as forest for which the trees are mainly standards produced through sexual reproduction from seedlings, and the term *even-aged* characterizes stands that consist of trees of a single class in which the range of age_{tree} is usually ±20% of the rotation age (Helms 1998) or stands for which age differences among individual trees are small, usually less than 20% of rotation age (http://www.iufro.org/science/special/silvavoc/silvaterm-database/). Thus, bridge building must focus on the comparability of stand age only. Actually, this lack of suitable data provides an opportunity to propose ecologically relevant new variables that could be surveyed in future NFIs with relatively low expense such as age_{dom} and age_{tree}. In the sections that follow, we propose an indicator for each of the three forest age variables.

For the age_{dom} variable, the proposed indicator is proportion of "old plots". In this context, old plots are defined as having age_{dom} based on ages of sample trees that are more than half the natural life span of the dominant trees species on the plot where the dominant tree species has the greatest basal area on the plot. We define the natural life span of a tree species to be the average number of years from germination to natural death of the 100 oldest (largest) trees per hectare in the upper canopy layer growing in humanly undisturbed natural or virgin forests in the terminal phase. Because such forests are quite rare in Europe, there are few reliable references on natural life spans according to our definition. A first draft of the average life spans of dominant tree species by European forest category (Table 5.2) is reported in Table 5.6. Many entries in Table 5.6 are estimates or refer to the maximum ages of solitary trees observed in non-forest conditions. These trees do not completely satisfy the criteria for trees used to estimate natural life span, and may have natural life spans longer than would be observed in virgin forests. Because this indicator might be difficult to report by dominant tree species and forest category, we propose instead a natural life span per forest category rather than for each species. Estimates could be based on the life span of the most dominant species of a specific forest category (Table 5.6).

For the age_{stand} variable, the proposed indicator is the proportion of stands older than 120 years in even-aged high forest. Because the NFIs of most countries currently assess age_{stand} only in even-aged high forest, we propose this indicator to guarantee a minimal comparison based on existing data. The threshold of 120 years is the maximal age assessed and reported by all countries except one and is proposed to be used as a bridge for facilitating harmonized international estimation. For most forest types, 120 years is still less than half the natural life span. Because the proportions of coppice forest, coppice with standards and uneven-aged forest differ considerably among countries, the utility of this indicator is greatly restricted at the international level.

For the age_{tree} variable the indicator is the mean number and proportion of old trees per hectare. An old tree is defined as having age greater than half the natural life span for its species in a specific forest type or forest category (Table 5.6). Because the ages of all sample trees on NFI plots are obtained by only very few countries, this indicator cannot be used yet for international comparisons. Thus, we propose that in the future countries either assess tree age in their inventories or develop bridges that permit estimation of tree age based on tree data such as

Table 5.6 Average natural life spans of dominant tree species for the 14 European forest categories (Table 5.2, EEA 2006). In bold the age of most common species in each forest category used for averaging

	Forest category													
Species	1	2	3	4	5	6	7	8	9	10	11	12	13	14
Abies alba	–	–	**400**	–	–	–	**425**	–	–	–	–	–	–	375
Acer campestre	–	–	–	–	**150**	–	–	100	–	–	–	–	–	–
Acer platanoides	–	–	–	150	150	–	350	150	–	–	–	–	–	–
Acer pseudoplatanus	–	–	–	–	400	–	350	–	–	120	–	–	–	–
Alnus glutinosa	–	–	–	–	110	–	–	–	–	–	110	–	**120**	–
Alnus incana	–	–	–	–	–	–	–	–	–	–	50	–	60	–
Betula pendula	150	100	–	**100**	120	–	–	–	–	–	**150**	100	**100**	–
Betula pubescens	100	60	–	60	–	–	60	–	–	–	60	–	60	–
Carpinus betulus	–	–	–	–	**150**	150	200	80	–	–	–	–	–	–
Castanea sativa	–	–	–	–	–	–	500	**500**	–	–	–	–	–	–
Fagus sylvatica	–	–	–	–	250	**250**	**280**	280	–	280	–	–	–	–
Fraxinus angustifolia	–	–	–	–	**250**	–	–	250	–	–	–	–	–	–
Fraxinus excelsior	–	250	–	250	**250**	250	250	**250**	–	–	–	**250**	–	–
Fraxinus ornus	–	–	–	–	–	–	100	100	–	–	–	–	–	–
Junglans regia	–	–	–	–	–	–	–	**200**	–	–	–	–	–	–
Larix decidua	–	–	**400**	–	–	–	–	–	–	–	–	–	–	–
Ostrya carpinifolia	–	–	–	–	–	–	150	200	–	–	–	–	–	–
Picea abiea	**300**	300	**450**	–	–	–	**250**	–	–	300	**250**	–	–	**250**
Pinus cembra	–	–	700	–	–	–	–	–	–	–	–	–	–	–
Pinus heldreichii	–	–	200	–	–	–	–	–	–	200	–	–	–	–
Pinus nigra	–	500	500	–	–	–	500	–	–	**500**	–	–	–	500
Pinus pinaster	–	–	–	–	–	–	–	–	–	**150**	–	–	–	–
Pinus pinea	–	–	–	–	–	–	–	–	–	**200**	–	–	–	–
Pinus sylvestris	**350**	**300**	250	–	–	–	250	250	–	**250**	–	–	–	**250**
Pinus uncinata	–	–	**300**	–	–	–	–		–	–	–	–	–	–

(continued)

Table 5.6 (continued)

Species	Forest category 1	2	3	4	5	6	7	8	9	10	11	12	13	14
Populus tremula	–	100	–	100	–	–	–	80	–	–	100	–	**100**	–
Prunus avium	–	–	–	–	80	–	–	–	55	–	–	–	–	150
Pseudotsuga menziesii	–	–	–	–	–	–	–	–	–	–	–	–	–	400
Quercus cerris	–	–	–	–	–	–	200	200	–	–	–	–	–	–
Quercus coccifera	–	–	–	–	–	–	–	–	**200**	–	–	–	–	–
Quercus frainetto	–	–	–	–	–	–	–	150	**150**	–	–	–	–	–
Quercus ilex	–	–	–	–	–	–	–	300	300	–	–	–	–	–
Quercus petraea	–	–	–	–	**550**	550	500	275	–	–	–	–	–	–
Quercus pubescens	–	–	–	–	225	–	–	**225**	**225**	–	–	–	–	–
Quercus robur	–	650	–	**650**	**500**	330	–	450	**450**	–	–	–	–	–
Quercus suber	–	–	–	–	–	–	–	250	**250**	–	–	–	–	–
Salix caprea	–	–	–	–	80	–	–	–	–	–	–	–	–	–
Salix spp.	–	–	–	–	–	–	–	–	–	–	150	**150**	–	–
Sorbus aria	–	–	–	–	–	–	100	–	–	–	–	–	–	–
Sorbus aucuparia	–	100	–	100	100	100	100	–	–	80	100	–	–	–
Sorbus domestica	–	–	–	–	–	–	350	200	–	–	–	–	–	–
Tilia cordata	–	300	–	300	**300**	300	–	200	–	–	–	–	–	–
Tilia platyphyllos	–	–	–	–	350	–	–	450	–	–	–	–	–	–
Ulmus glabra	–	450	–	450	450	450	–	–	–	–	–	–	–	–
Ulmus minor	–	–	–	–	400	–	–	–	–	–	–	–	–	–
General proposal[a]	**330**	**300**	**380**	**400**	**320**	**260**	**320**	**260**	**250**	**280**	**200**	**200**	**110**	**250**

[a] Estimation based on values and proportions of the most dominant tree species growing in the European Forest Categories

species, diameter at-breast-height (dbh), height, and social position, and site and stand data such as altitude, soil, and forest type.

5.4.3 Components to Be Tested

Several questions regarding tree age were addressed using data from the common database:

(a) What is the influence of different national definitions on the comparability of age_{stand}?
(b) How much do estimated plot ages depend on the dbh measurement thresholds?
(c) What is the correlation between age_{tree} and dbh?
(d) Could bridges in the form of models be used to predict missing age_{tree}?
(e) Are the three proposed indicators practicable and do they produce comparable results?

5.4.3.1 Question (a), The Effect of Different Age_{stand} Definitions

As shown in Sect. 3.4, most national definitions of age_{stand} refer only to the dominant stand elements of trees, species, and layers. Few countries estimate age_{stand} as the weighted or unweighted average of the ages of all living trees in a stand. Thus, one question is whether data on age_{stand} reported by countries are comparable. Using tree data from the Swedish NFI, different definitions were applied and compared for three forest categories (Fig. 5.11). Four variables related to forest age are compared:

1. Age_{stand}: age of stands as reported by countries,
2. Age_{dom}: mean age of the 100 trees per hectare with the largest diameters calculated using the diameters and ages of the plot sample trees,
3. $Age_{weighted}$: basal area weighted mean of the ages of all plot sample trees,
4. Age_{mean}: mean age of all plot sample trees.

The results using data in the common database for Sweden show that age_{dom}, $age_{weighted}$, and age_{mean} are in general greater than age_{stand}, more or less independently of forest category (Fig. 5.11). Age_{dom} is the greatest followed by $age_{weighted}$ and finally by age_{mean}. The estimates for the forest age variables differ from 8% to 20%, depending on forest category. The differences between age_{stand} and age_{dom} are even greater (18–51%). Thus, the effects of using different definitions or methods may be that countries with quite similar forest age structures report quite different estimates. For example, two countries A and B report mean stand age of 70 and 76 years, respectively for hemiboreal and nemoral forests (forest category 2) (Fig. 5.12). This small difference of six years can be attributed to country A estimating stand age using ages of dominant trees and country B using the mean of all trees as the estimate. If the two countries used the same definition, the difference between the countries would be greater and would range from 10 to 25 years, depending on the age variable

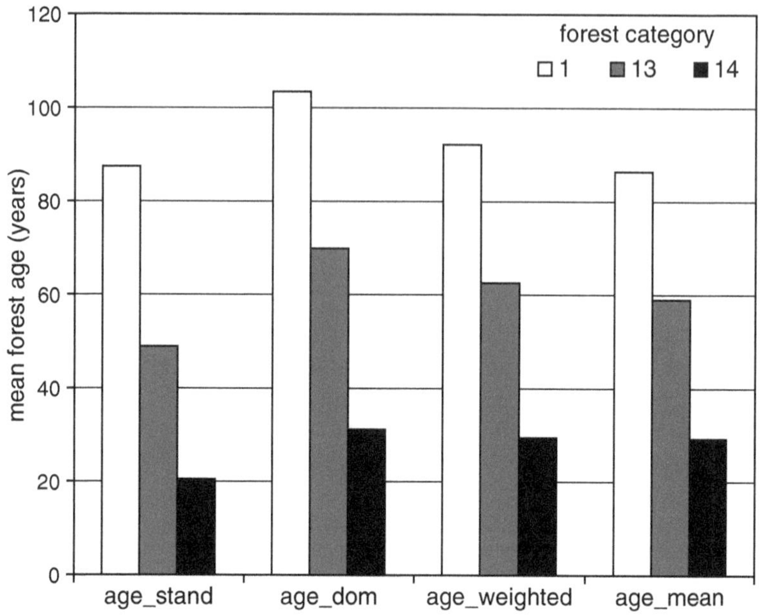

Fig. 5.11 Mean forest age by variable and forest category for Sweden

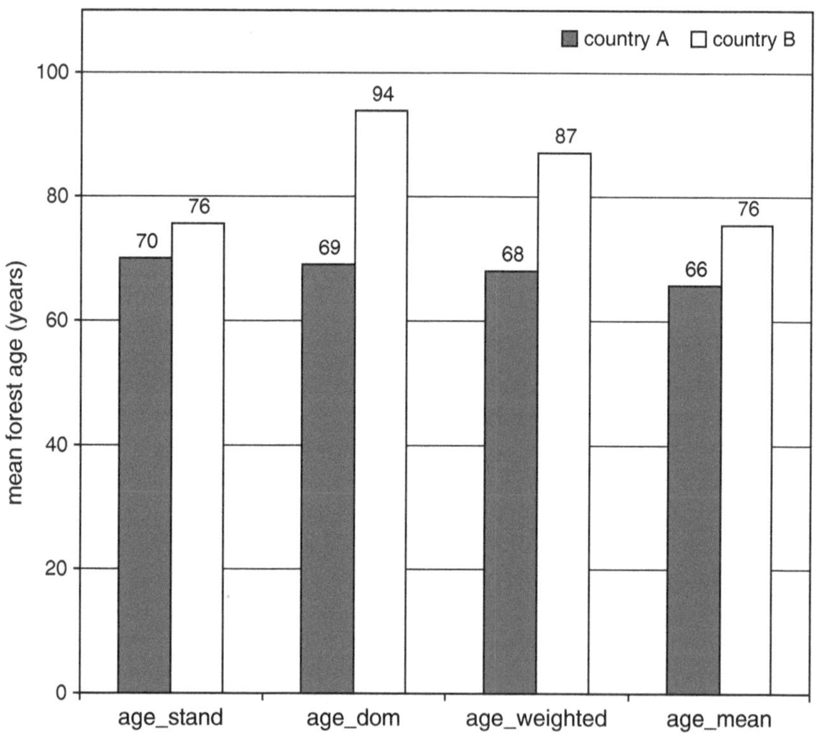

Fig. 5.12 Mean forest age by variable and country for the hemiboreal and nemoral forest category (forest category 2, Table 5.2)

used. Comparable effects result for other countries and forest categories when using NFI data and confirm the assertion that international comparisons of stand age results based on existing data are problematic and not recommended.

5.4.3.2 Question (b), Influence of dbh Measurement Threshold on Forest Age

We propose to estimate forest age using sample tree ages if possible, especially in uneven-aged stands. The effects of different dbh measurement thresholds used by the NFIs of European countries on estimated plot ages are tested with data from Sweden. The analyses attest that $age_{weighted}$ and especially age_{mean} depend on dbh measurement thresholds whereas age_{dom} does not show differences in median of forest age (Fig. 5.13). The variable, age_{dom} seems to be the most suitable for estimating comparable forest ages independently of dbh measurement thresholds. However, because the ages of all plot sample trees are used, the distribution of forest age as well as the minimum and maximum of estimates for all plot age variables may be different than age_{stand} because of remnants of former old stands and/or trees from adjacent stands (Fig. 5.13). With regard to the forest age indicators, these factors contribute to quite different results. The proportion of forests older than 120 years is remarkably greater for $age_{weighted}$ and age_{dom} than for age_{stand} (Fig. 5.14). Estimates for age_{mean} depend considerably on dbh

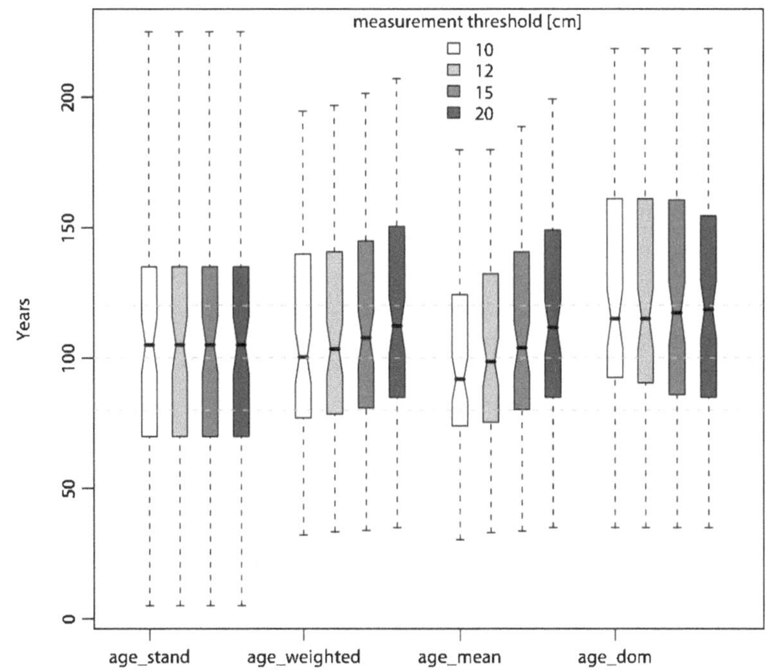

Fig. 5.13 Mean forest age by age variable and dbh measurement threshold (10, 12, 15, 20 cm) for Swedish NFI data

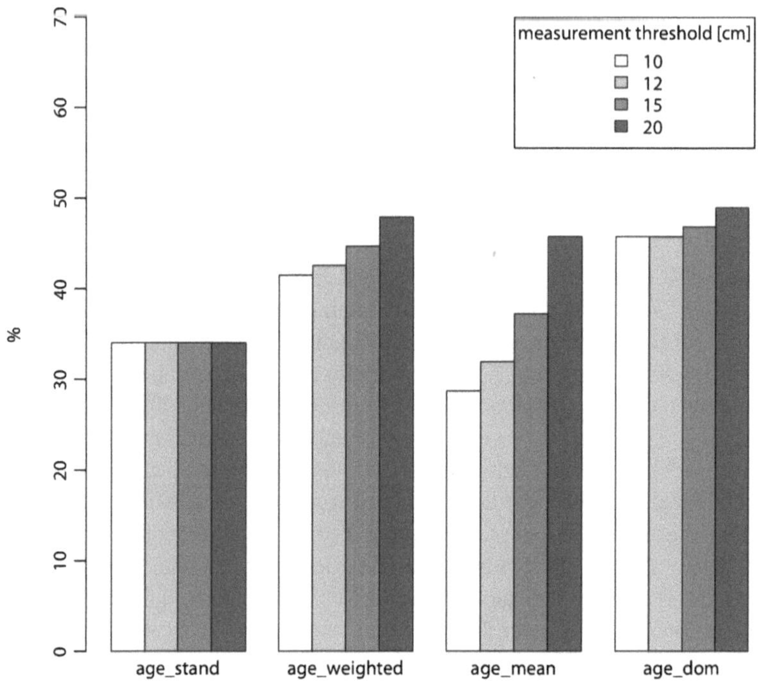

Fig. 5.14 Percentages of plots with forest age greater than 120 years by age variable and dbh measurement threshold using Swedish NFI data

measurement threshold whereas those for age_{dom} tend to be stable for small dbh thresholds up to 12 cm which is the greatest used in Europe (Switzerland and Cyprus).

5.4.3.3 Question (c), Correlation Between Age and Diameter

Because age_{tree} is currently assessed by only a few NFIs, a general question concerning indicators related to biodiversity is whether forest age can be replaced by an indicator based on dbh which is measured by all NFIs. However, Based on the Swedish NFI data, only 33% of tree age variation can be explained by tree dbh (Fig. 5.15). Thus, the variable age_{tree} cannot be replaced with an indicator based on dbh. At plot level, the strength of the relationship between age_{dom} and dbh_{dom} is much greater (R^2=0.54, Fig. 5.16). Thus, dbh_{dom} could be considered as an alternative for age_{dom} when tree age is not assessed or when tree age can be predicted using tree age-dbh models.

5.4.3.4 Question (d), Tree Age Models

If tree age is not assessed in the field by all NFIs, then age_{dom} and age_{stand} cannot be obtained directly and are not comparable because their definitions are not applied. One solution would be to develop models for predicting age_{tree} for different tree species

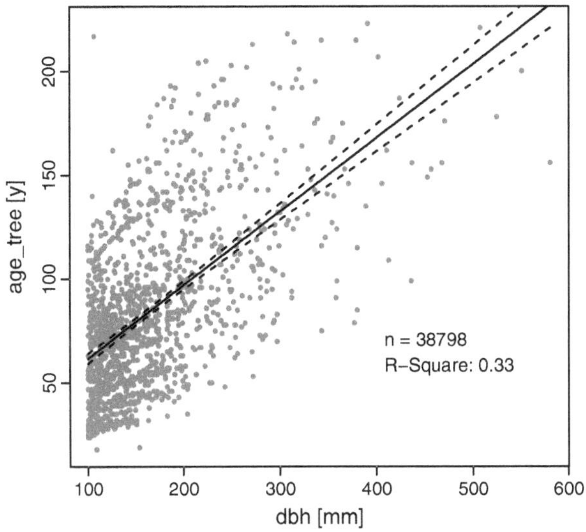

Fig. 5.15 Simple linear regression model of relationship between age_{tree} and tree dbh for Swedish NFI data

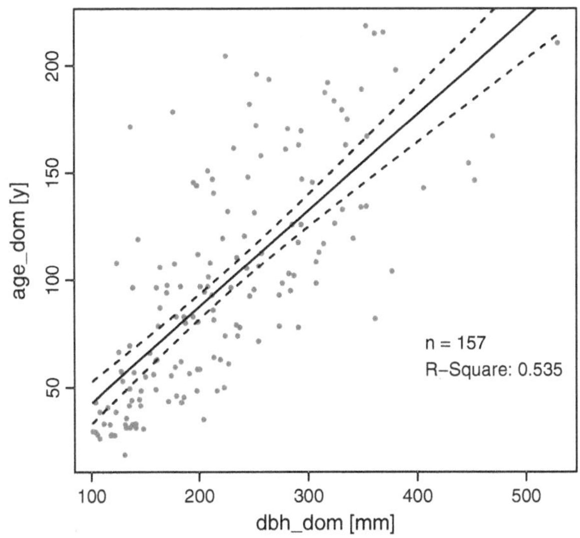

Fig. 5.16 Simple linear regression model of relationship between age_{tree} and dbh_{dom} for Swedish NFI data

and forest categories using existing data from NFIs that assess age_{tree} or research data obtained by each country. Analyses based on all age_{tree} data in the common database show that 37–55% of the variation in age_{tree} is explained by a combination of tree species, dbh and forest category data (Fig. 5.17). Similar strong relationships ($0.36 \leq R^2 \leq 0.55$) can be shown between true age and dbh of cut spruce trees using Swiss NFI data for four site classes (Fig. 5.18). When all relevant variables including

a

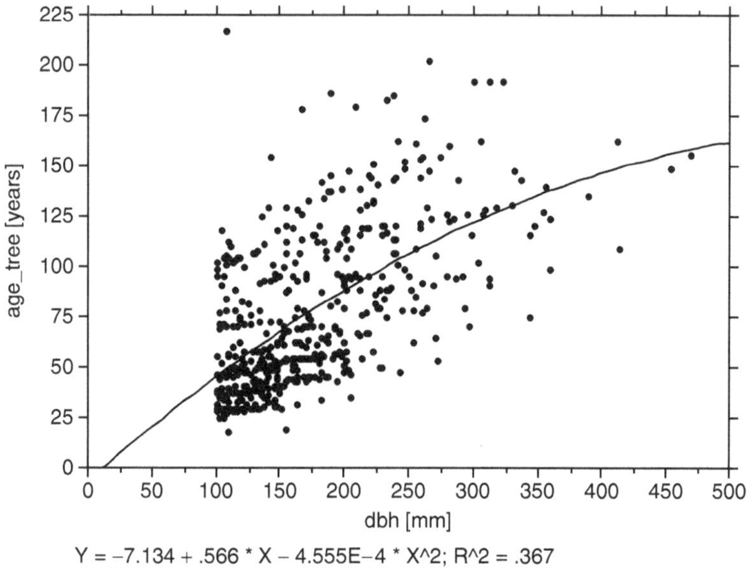

b

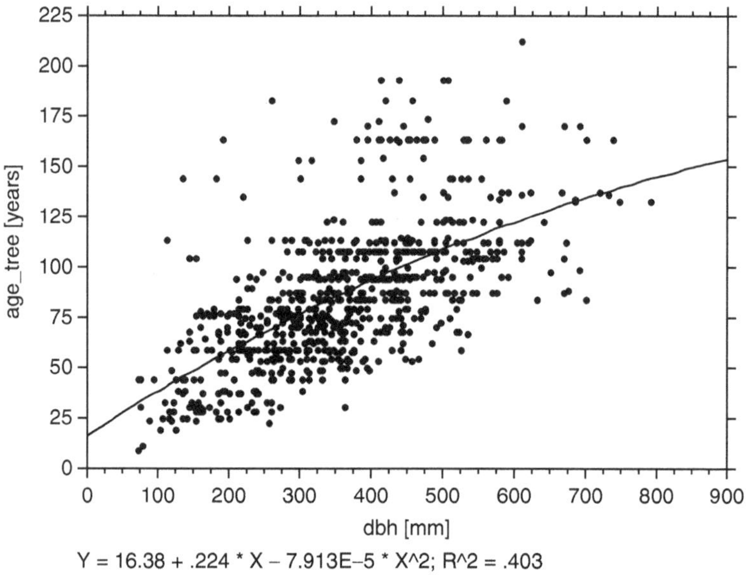

Fig. 5.17 Polynomial regression models for tree age versus dbh by tree species and forest category. Inclusion criteria: (**a**): Boreal forest form *Pinus sylvestris*, (**b**): Alpine coniferous forest from *Picea abies*, (**c**). Hemiboreal + nemoral forest from *Fagus sylvatica*, and (**d**): Mesophytic deciduous forest from *Quercus* spp

c

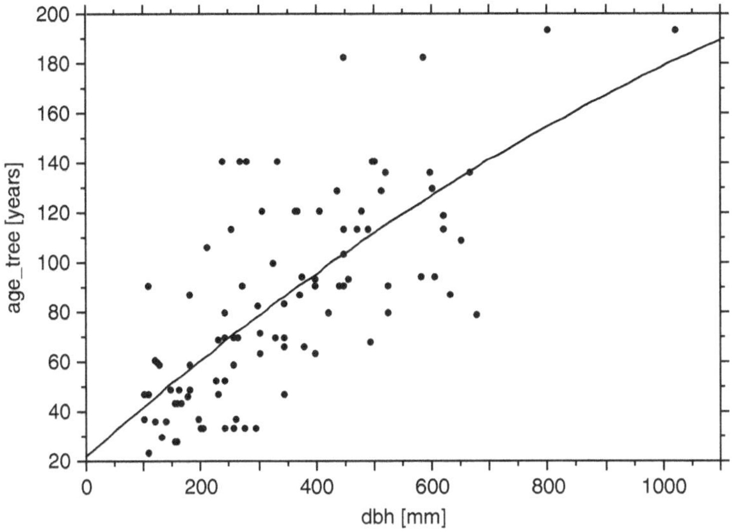

Y = 21.943 + .201 * X – 4.472E–5 * X^2; R^2 = .503

d

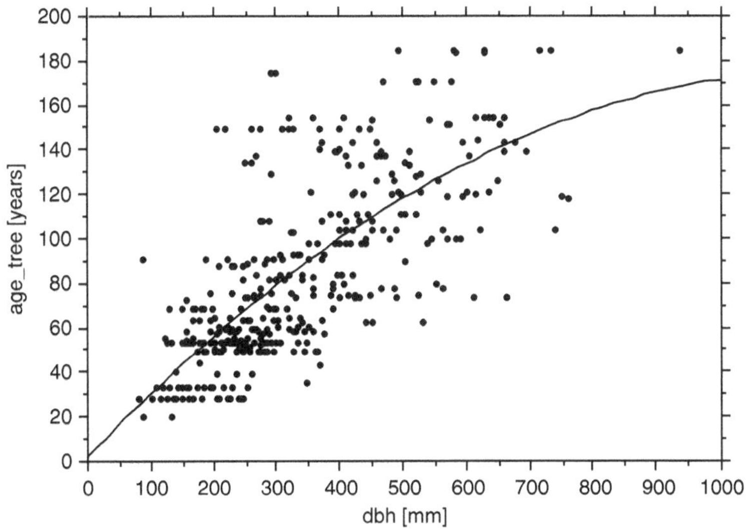

Y = 1.755 + .295 * X – 1.252E–4 * X^2; R^2 = .545

Fig. 5.17 (continued)

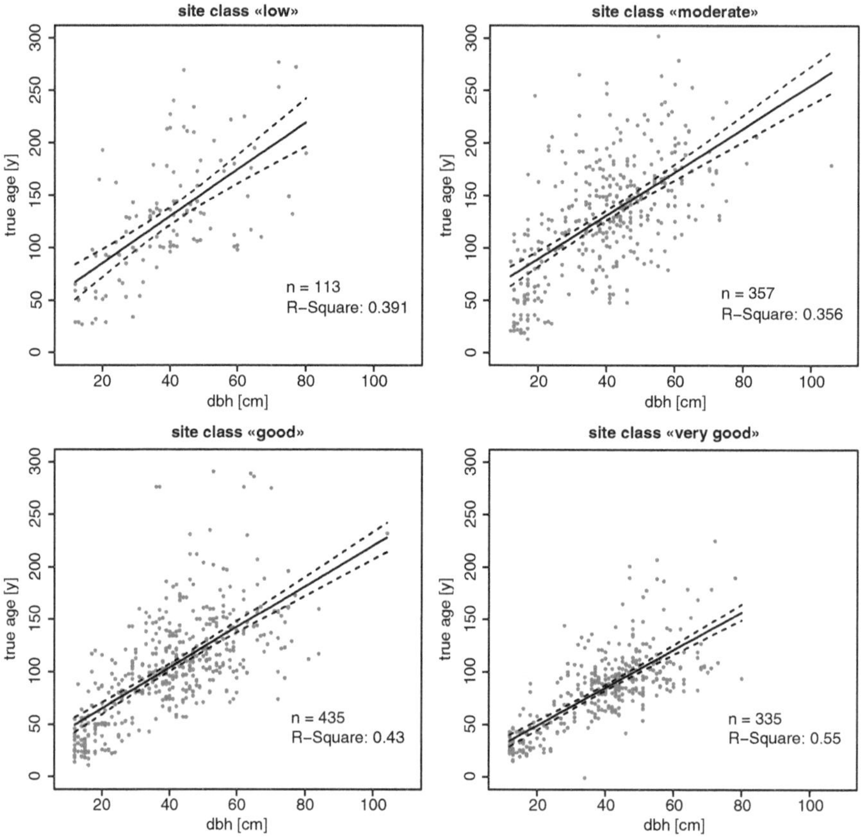

Fig. 5.18 Simple linear regression model of relationship between true tree age and dbh of cut spruce by site class using Swiss NFI data

site, tree and stand factors are considered, models could produce even stronger correlations. However, first results for a case study in Switzerland show that models can produce comparable or even better predictions of true tree age than in situ estimates by field teams, apart from drilling (Abegg, personal communications). To acquire as much data as possible for developing models, the true ages of cut sample plot trees are determined in the Swiss NFI by counting stump tree rings in following surveys. A first simple model to predict tree age of spruce was developed for the relatively homogeneous Swiss region, Plateau east, using the average of true age by 5 cm dbh classes (Brändli, personal communication) (Fig. 5.19).

5.4.3.5 Question (e), Practicability of the Proposed Forest Age Indicators

In Sect. 5.4.2, an indicator for each of the forest age variables was proposed: for age_{dom}, the proposed indicator is proportion of old plots; for age_{stand}, the proposed

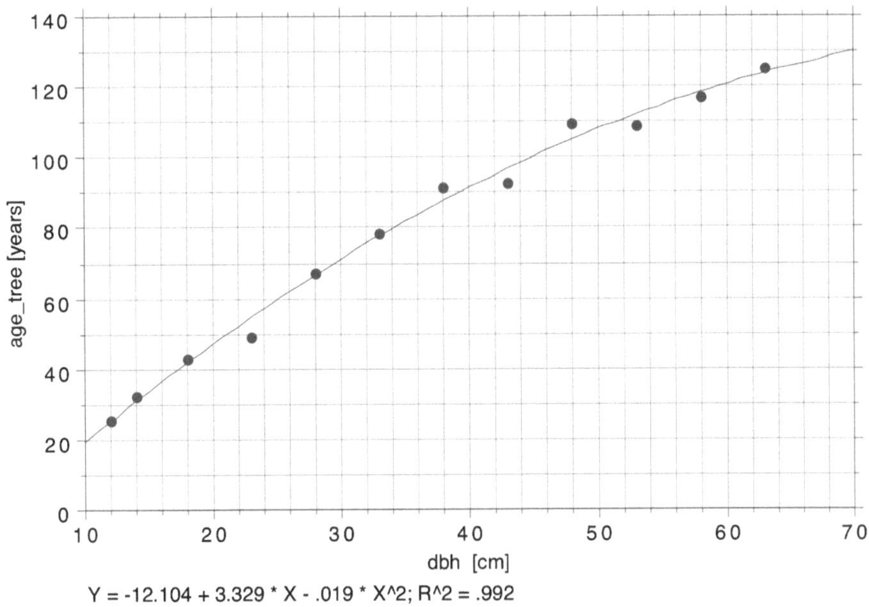

Fig. 5.19 NFI tree age model for spruce in the Swiss region, Plateau east

indicator is the proportion of stands older than 120 years in even-aged high forest; and for age_{tree}, the proposed indicator is the mean number and proportion of old trees per hectare. These indicators are analysed by country and forest category.

Proportions of old plots (indicator 1, age_{dom}) were estimated by forest categories (Table 5.6) using age_{dom} based on the ages of all trees on plots relative to the natural life spans of the dominant species (Table 5.5). In general, relatively few plots have age_{dom} that is greater than half the natural life span of the dominant tree species. Estimates differ from 0.00 to 0.33 depending on forest category and country (Fig. 5.20). For the Plantation forest category (forest category 14), age_{dom} did not reach half the natural life span of the dominant tree species on any plot, and estimates for the hemiboreal and nemoral forest (forest category 2) were very small, also. However, such is not the case for the Beech forest (forest category 6).

Proportions of stands older than 120 years (indicator 2, age_{stand}) show large differences among countries and forest categories (Fig. 5.21). Countries G and I have no old stands, whereas countries A and E have proportions of 0.45 and 0.50. The ranking among countries differs depending on the indicator. In Mesophytic deciduous forest (forest category 5), countries A and B have the same estimates for this indicator, whereas estimates for proportion of old plots (indicator 1) differ considerably.

Proportion of trees older than half their natural life spans (indicator 3, age_{tree}) is practicable with results comparable to those for proportion of old plots (indicator 1) with estimates between 0.00 and 0.47 depending on forest category and country. Further analyses indicated that greater thresholds (e.g., proportion of trees older

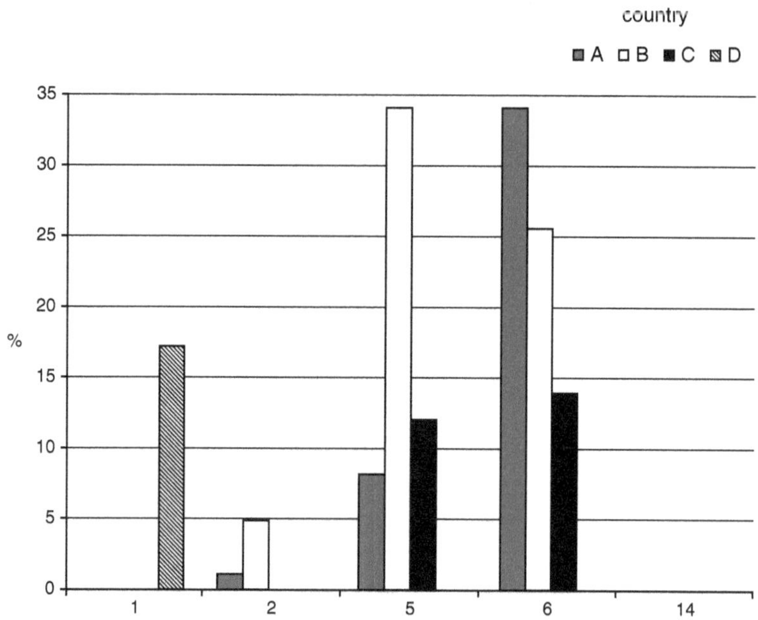

Fig. 5.20 Proportion of plots older than half the natural life span of their dominant tree species by country and forest category

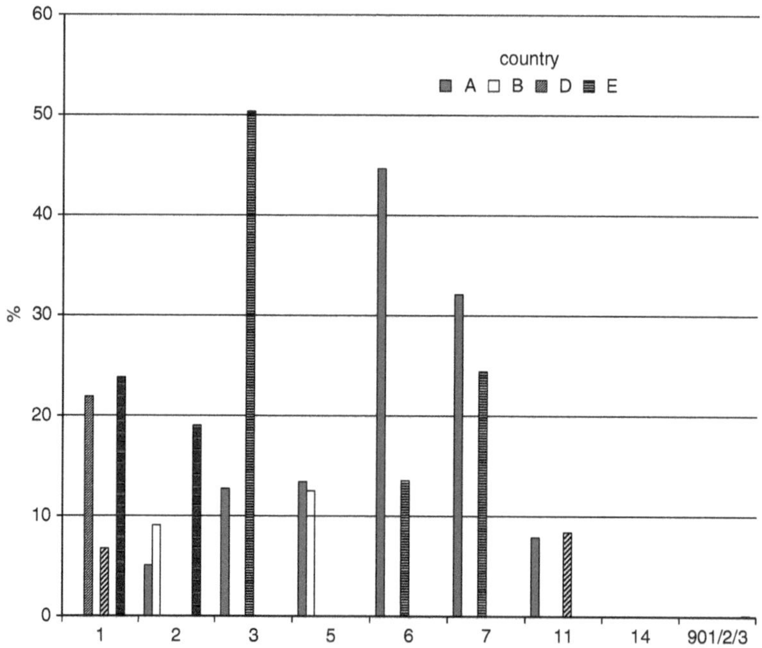

Fig. 5.21 Proportion of stands older than 120 years in even-aged high forest by country and forest category

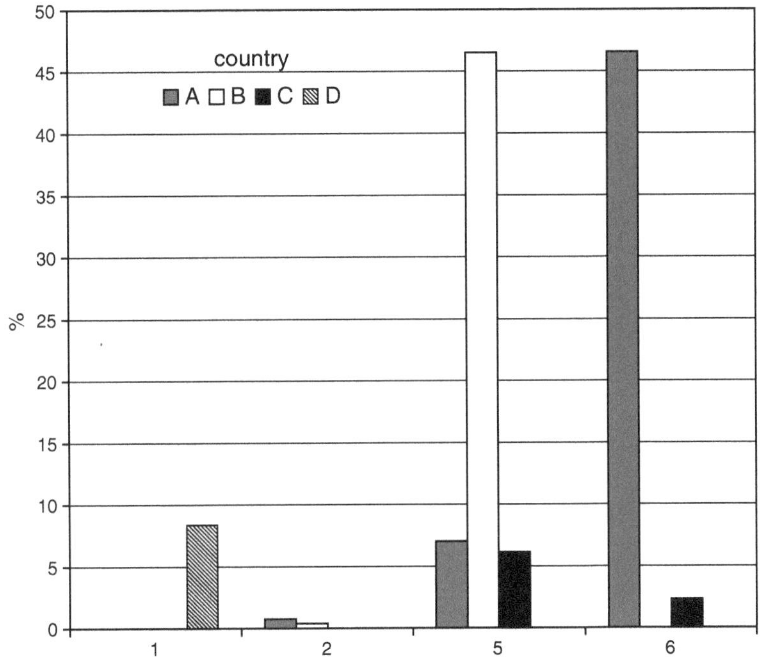

Fig. 5.22 Proportion of trees older than half of their natural life span by country (A–D) and forest category (1, 2, 5 and 6)

than 75% or 90% of their natural life spans) are not practicable because the proportions for most countries and forest categories are 0% and therefore not useful for comparison purposes.

In summary, all three indicators are practicable. Proportion of old plots (indicator 1, age_{dom}) and proportion of trees older than half their natural life spans (indicator 3, age_{tree}) produce similar rankings among countries within the same forest category. However, estimates for these two indicators are not comparable to estimates for proportion of stands older than 120 years (indicator 2, age_{stand}). Thus only one of the three forest age variables should be selected for international reporting (Fig. 5.22).

5.4.4 Discussion

The existing data on age_{stand} are not comparable because most countries estimate age_{stand} only in even-aged high forest. However, for purposes of assessing forest biodiversity, all stands must be considered, not just those in even-aged high forest. Thus, new approaches to assessing stand age are necessary.

National definitions of age_{tree} and age_{stand} are in general quite similar, and there is a consensus that the dominant trees of a stand or plot should be used to estimate forest age. The few countries that use the unweighted mean of the ages of all plot sample trees are encouraged to change their definitions. We propose using the definition of age_{dom} because this variable can be used in both even-aged and uneven-aged forests. Further, the variable is mostly independent of the dbh measurement threshold and is the most relevant with respect to biodiversity assessments.

Estimation of age_{dom} requires the ages of the dominant trees; thus, this variable should be assessed by NFIs. The Swiss NFI has begun to assess the ages of all plot trees in its new NFI that started in 2009. In young growth and thickets without sample trees, age_{dom} could be estimated. An alternative to estimating age_{tree} in the field could be to use models based on existing NFI data or data from research plots to predict age_{tree}. If age_{tree} is not assessed in NFIs or is predicted using tree age models, dbh_{dom} could be considered a surrogate or alternative for age_{dom}.

Because the results for the three forest age variables are not directly comparable, we recommend using only one, age_{dom}. The rationale for the recommendation is threefold: age_{dom} is the most relevant for biodiversity assessment purposes, age_{tand} is only partially available, and age_{tree} cannot generally be replaced by dbh.

5.5 Deadwood

5.5.1 Introduction

The deadwood harmonization test used raw deadwood data from nine European countries and the United States of America (USA). The common database (Chap. 4) included data for 9,267 plots of which 5,012 had positive deadwood observations. The test addressed two main objectives: to analyze the feasibility of deadwood harmonization procedures, and to evaluate the impact of different minimum diameter thresholds on final deadwood estimates. Harmonization methods were investigated for estimates of deadwood volume per unit area, spatial position, decay class and species group.

5.5.2 Materials

The common database included raw data for 4,901 plots from nine European countries and for 4,366 plots from the USA; the data represented 23,607 observed and measured deadwood elements. The European plots were distributed as follows: 400 plots (8%) from the Walloon region of Belgium, 401 (8%) from Switzerland, 302 (6%) from the Czech Republic, 790 (16%) from Germany, 1,458 (30%) from Denmark, 775 (16%) from Spain, 336 (7%) from Finland, 251 (5%) from Italy, and 188 (4%) from Sweden. Each country was responsible for selection of the plots to be included in the common database. All countries measured diameter at-breast-height (dbh) and height for standing deadwood elements, whereas measures for lying deadwood were for whole deadwood pieces for some countries but only for portions for other countries.

To harmonize estimates of deadwood based on proposed reference definitions, a complete understanding of local methods and definitions adopted by the ten countries was needed for several deadwood topics including sampling methods, spatial position, decay classes, woody species, and volume estimation.

5.5.3 Methods

Three types of bridges have been developed on the basis of the relationships between local and reference definitions (Ståhl et al. submitted, Vidal et al. 2008): *neutral* bridges for which country data correspond to the reference definition, *expansive* bridges for which country data are more restricted than the reference definitions in which case additional information is necessary, and *reductive* bridges for which country data exceed the reference definition requirements. Bridges were used at two different levels: estimation of volume for individual pieces of deadwood, and estimation of per plot volume by categories of spatial position, decay class, woody species. Reductive and neutral bridges were used for single pieces of deadwood. Single-piece estimates were then aggregated at plot level and expansive bridges were constructed and used where needed. The procedure was repeated for three different reference definitions corresponding to three minimum diameter thresholds: 10 cm, 12 cm, and 20 cm, labelled respectively Ref_{10}, Ref_{12} and Ref_{20}. The different thresholds were used to evaluate the impact of deadwood definitions on the final aggregated estimates.

5.5.3.1 Spatial Position

Two countries use complex systems of nomenclature based on four or five classes to describe the spatial position of deadwood elements, whereas all the other countries use simpler systems based on two classes which were used as references, lying and standing (Table 5.7). Reductive bridges were used for two countries and neutral bridges for the other countries.

5.5.3.2 Decay Classes

Information on deadwood decomposition stage was available for nine countries all of which used local systems of nomenclatures with numbers of decay classes ranging from three to nine based on deadwood colour, texture and softness. Bridges were developed to reclassify the local decay classes into five reference classes following a gradient of increasing firm texture: A, B, C, D and not available (Table 5.7). Neutral bridges were necessary for three countries, whereas, reductive bridges were used for the other countries.

Table 5.7 Deadwood reference definitions adopted for the harmonization test

Reference number	Deadwood elements	Reference definitions
1	Living and dead stems	A living stem has active or dormant cambium; otherwise the stem is dead
2	Standing and lying stems	A lying stem is the main stem which is not self-supporting with the majority of its length lying on the ground; otherwise it is a standing stem
3	Decay classes	Four decay classes (A, B, C, D) are considered on the basis of the percent of hard texture wood present in the deadwood volume. Wood is considered "hard texture" if a knife cannot be penetrated more than 2 cm – Class A: hard texture ≥90% (not decayed, completely hard) – Class B: hard texture 90–60% (slightly decayed, most part still hard) – Class C: hard texture 60–30% (decayed, most part soft) – Class D: hard texture ≤30–5% (very decayed, completely soft)
4	Stem volume of dead trees	The stem volume of dead trees is the aggregated above-ground volume of all dead stems, standing or lying, over a specified area. Included are over-bark stem volumes – from the stump height to a top over-bark diameter of 10 cm – of dead stems with a diameter at breast height of more than 10 cm. Branches are excluded
5	Piece of coarse woody debris	A piece of coarse woody debris is a downed (not suspended) piece of deadwood lying on ground, with sections coarser than 10 cm (over bark) of at least 1 m in length. Lying dead stems, including attached branches, are excluded
6	Volume of coarse woody debris	The volume of coarse woody debris is the aggregated above ground volume of all pieces of coarse woody debris over a specified land area. Included are over-bark volumes of those sections of the coarse woody debris pieces, which are coarser than 10 cm (over bark) on a length of at least 1 m

5.5.3.3 Woody Species

Six countries assessed the species of the deadwood elements, whereas the other countries recorded species only if the element was from a coniferous or broadleaved tree. In some cases, countries used *unidentified* to denote advanced levels of decomposition. Harmonization was carried out to classify all deadwood pieces represented in the common database into three classes: coniferous, broadleaved, not available.

5.5.3.4 Volume Estimation

Harmonization of deadwood volume estimates was conducted in two steps. First, harmonized estimates of volume were calculated for each deadwood piece represented in the common database using either neutral or reductive bridges. The process was repeated for each threshold: Ref_{10}, Ref_{12} and Ref_{20}. Estimates of deadwood volumes for individual pieces were then aggregated for each NFI plot. Second, when needed, expansive bridges were used to estimate plot-level volume because of differences in both minimum diameter and minimum length thresholds. For Ref_{10}, expansive bridges were necessary for six countries; for Ref_{12}, expansive bridges were necessary for five countries; and for Ref_{20}, expansive bridges were necessary for four countries.

5.5.3.5 Volume Estimation: Per Piece Harmonization

Bridges were developed to estimate volume corresponding to the reference definition for each individual deadwood piece represented in the database. A reductive bridge was used when local country thresholds were less than that for the reference thresholds. All reductive bridges were expressed as a reducing factor (*Rf*) for each deadwood piece. The *Rf* values were between 0 and 1 and were used to reduce the piecewise deadwood volume estimate provided by the countries. Calculation of *Rf* for the per piece harmonization depended on the deadwood component: dead stems or dead coarse woody debris (cwd).

5.5.3.6 Volume Estimation: Per Piece Harmonization for Standing Dead Stems

The reference definitions specify that standing dead stem volume includes stem volumes from the stump height to a top over-bark diameter of 10 cm for stems with dbh of at least 10, 12, or 20 cm. The local minimum dbh can be smaller than, equal to, or greater than the minimum dbh in the reference definition. When they are equal, no or neutral harmonization is necessary. When the local minimum dbh is smaller than the minimum dbh of the reference, the bridge is reductive, and when the local dbh is greater than the minimum dbh of the reference definition, an expansive bridge is required.

Some countries use local stem volume definitions that include minimum top diameters that can be equal to or smaller than 10 cm. Stem volume from the stump height to a reference top diameter of 10 cm ($Vtop_{ref}$) can be estimated using the following relationship (Corona and Ferrara 1992),

$$Vtop_d = Rf_d * Vtop_0, \tag{5.1}$$

where $Vtop_0$=total stem volume, $Vtop_d$=stem volume to top diameter d, Rf_d=reduction factor. Rf_d can be calculated from a simplified model developed by Corona and Ferrara (1992),

$$Rf_d = 1 - \left[(H - 1.3)/H\right]^{2b+1} \times \left(d / dbh^{(2b+1)/b}\right), \tag{5.2}$$

where H is the total height of the stem, b is a parameter estimated for each tree, and d is top diameter. The stem volume to a reference top diameter is then defined as follows,

$$Vtop_{ref} = Rf_{ref-d} * Vtop_d, \tag{5.3}$$

where: *ref*=reference based on a minimum threshold of 10 cm, $Vtop_{ref}$ = reference stem volume from the stump height to a top diameter of 10 cm, $Vtop_d$=stem volume from stump height to a top diameter d, and Rf_{ref-d}=reduction factor from $Vtop_d$ to $Vtop_{ref}$.

5.5.3.7 Volume Estimation: Per Piece Harmonization for Dead Coarse Woody Debris

The shape of dead cwd was assumed to be the frustum of a cone defined by maximum diameter (D_{max}), minimum diameter (D_{min}), length (L) as the linear distance between D_{max} and D_{min}, and median diameter (D_{median}) as the diameter measured at half-length. The tapering rate (R) is then,

$$R = \left(D_{max} - D_{min}\right)/L. \tag{5.4}$$

The volume of cwd, according to the adopted reference definitions, is the volume of the portions of a piece of lying deadwood with a minimum diameter equal to or greater than 10 cm (or 12 or 20 cm) and having at least 1 m length. Pieces of deadwood were ideally divided into two components: one with the minimum diameter (and length) larger than the reference thresholds (its volume is $V_{Dmin\ ref}$) and one with the minimum diameter (and/or length) smaller than the reference thresholds (its volume is $V_{Dmin<ref}$). Calculation of $V_{Dmin\ ref}$ and $V_{Dmin<ref}$ is carried out on the basis of the tapering rate R defined with different methods depending on the country data available (D_{max} and D_{min} or D_{median}). Additional details are provided by Rondeux et al. (submitted).

5.5.3.8 Volume Estimation: Per Plot Harmonization

Per plot harmonization was necessary for countries whose minimum diameter or the minimum height/length used in the selection of measured deadwood elements were greater than the reference definition thresholds. In these cases, expansive bridges are needed to estimate the portion of the deadwood element not measured in the field using the national definitions and consequentially not available in the common database. In the other countries where expansive bridges were not needed, the final harmonized estimate of deadwood volume per plot is the sum of the per piece harmonized volumes of all the deadwood pieces available in the database for the plot, expressed on a per hectare basis.

Harmonization was carried out at the plot level by constructing a model of the relationship between the deadwood volume estimated using the minimum thresholds used by the countries (V_{NFI}) and the volume estimated using the reference definitions (V_{ref}). For a sample of plots from the Czech Republic, Italy, and Spain available in the database, the relationship between V_{NFI} and V_{ref} was described using a simple linear model of the form,

$$V_{ref} = a * V_{NFI} + b, \tag{5.5}$$

which was found to be adequate and was used as an expansive per plot bridge. Additional details are provided by Rondeux et al. (submitted).

5.5.4 Results

Final harmonized estimates of deadwood volume are reported for forest categories, spatial positions, decay classes and woody species. For the 5,012 plots available in the common database where deadwood volume follows local definitions (V_{NFI})>0, the average harmonized estimate of deadwood volume decreases as the minimum diameter threshold increases: moving from V_{NFI} to V_{Ref10}, V_{Ref12} and V_{Ref20}, the average deadwood volume estimate varies from 15.84 (SE=0.570), to 15.33 (SE=0.593) to 14.66 (SE=0.583) to 11.17 m^3ha^{-1} (SE=0.548).

For all 9,267 plots, which includes the 4,255 plots for which no deadwood was observed and for which deadwood volume is zero, regardless of reference definition, harmonized estimates of deadwood volume for V_{NFI}, V_{Ref10}, V_{Ref12} and V_{Ref20} decreased respectively from 8.56 (SE=0.319) to 8.28 (SE=0.330) to 7.91 (SE=0.324) and to 6.03 m^3ha^{-1} (SE=0.302). The ratio of estimates of lying deadwood to standing deadwood remains quite stable moving from V_{NFI} to V_{Ref10}, V_{Ref12} and V_{Ref20}; in particular, standing deadwood volume estimates represent 71%, 73%, 73% and 74%, respectively, of the total deadwood volume estimate. Similarly, percentages of estimated total deadwood volume represented by estimates for conifers, broadleaves and unclassified are 55%–40%–5%, 57%–39%–4%, 57%–40%–4% and to 56%–41%–3% for V_{NFI}, V_{Ref10}, V_{Ref12} and V_{Ref20} respectively.

The distribution of deadwood volume estimates for the four harmonized decay classes is quite stable, independently of the reference definition. Estimates of percentages of total deadwood volume for the five classes (A, B, C, D, not available) were 16%–21%–44%–11%–8% for V_{NFI}; 15%–21%–46%–10%–8% for V_{Ref10}; 15%–21%–46%–10%–8% for V_{Ref12}; and 15%–21%–46%–10%–8% for V_{Ref20}.

For all 14 European forest categories (Table 5.2), deadwood volume estimates decreased when changing from the NFI definitions to the three reference definitions (Fig. 5.23). The percentage reductions in deadwood volume estimates when changing from V_{NFI} to V_{Ref10}, V_{Ref12}, V_{Ref20} respectively were least for Beech forests (forest category 6) (2.4%, 4.8%, 10% respectively) and the greatest for Alpine coniferous forests (forest category 3) (32.5%, 40.7%, 71.2% respectively).

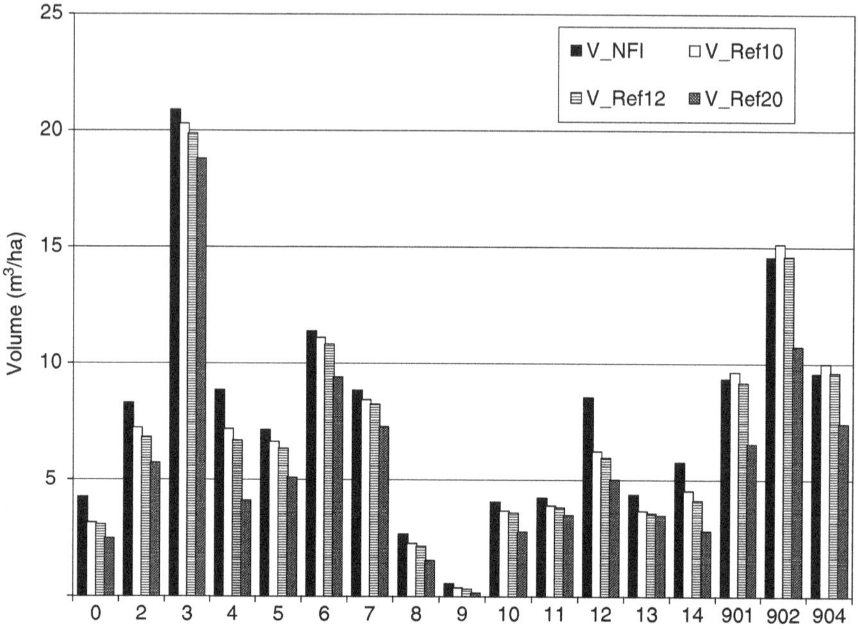

Fig. 5.23 Mean volume by forest category for 9,267 plots using local definitions (V_{NFI}) and reference definitions (V_{Ref10}, V_{Ref12} and V_{Ref20}). (Table 5.2 for European forest category names; American forest types: 901-aspen, 902-paper birch, 904-balsam poplar)

For the three American forest types, changing from V_{NFI} to V_{Ref10} produced an increase in mean deadwood volume estimates of 3.2% for aspen (forest type 901), 3.8% for paper birch (forest type 902) and 4.5% for balsam poplar (Forest Type 904), and changing from V_{NFI} to V_{Ref12} had a limited impact (–1.6%, –0.1% and –0.4% respectively) for the three forest categories. As expected, changing from V_{NFI} to V_{Ref20} produced a decrease in mean deadwood volume estimates ranging from 22.5% for balsam poplar to 29.8% for aspen. The mean deadwood volume estimates reported correspond only to the plots represented in the common database and should not be construed to be representative of the entire USA.

The impacts on mean deadwood volume estimates when changing from V_{NFI} to V_{Ref10} vary by country (Fig. 5.24) from a 3.3% increase (country 5) to a 30.3% decrease (country 1). Changing from V_{NFI} to V_{Ref12} produced a decrease in mean deadwood volume estimate by country ranging from 1.1% (country 1) to 40.5% (country 5). Finally, changing from V_{NFI} to V_{Ref20} produced decreases ranging from 9.8% (country 10) to 63.5% (country 1).

5.5.5 Discussion

The test focused on constructing bridges to produce harmonized estimates based on reference definitions using data collected according to national definitions.

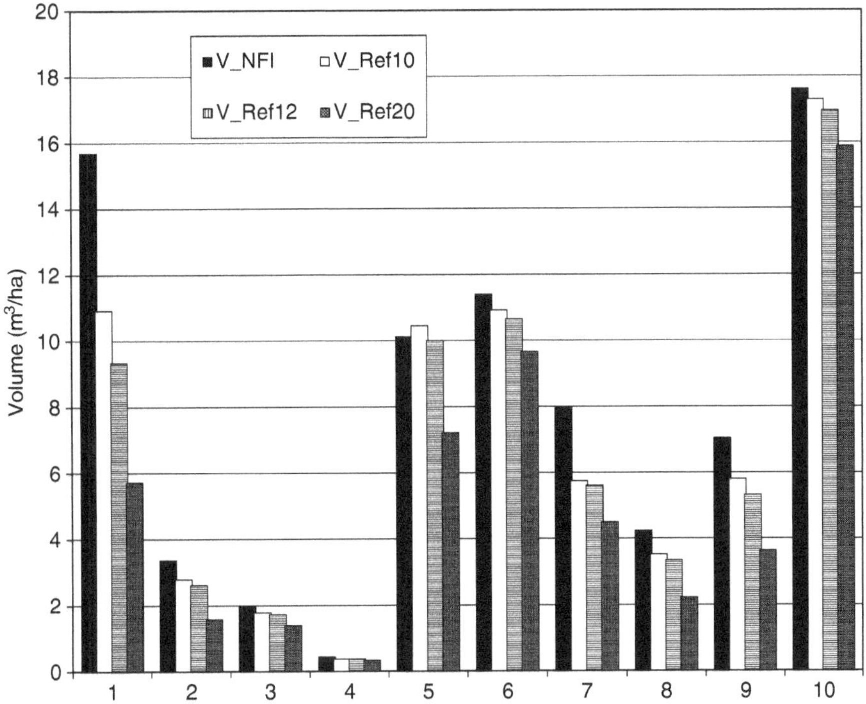

Fig. 5.24 Mean volume by country for 9,267 plots using NFI definitions (V_{NFI}) and reference definitions (V_{Ref10}, V_{Ref12} and V_{Ref20}); country codes are randomly assigned

The harmonization process is easier when local country definitions of deadwood pieces have minimum thresholds (in terms of diameter and length) equal to or smaller than the thresholds of the international references definitions. In these cases, neutral or reductive bridges can be used. However, when local minimum thresholds are greater than the reference thresholds, more complex expansive bridges are necessary because additional information is necessary.

Several primary conclusions may be drawn from this study. First, and most importantly, the results clearly indicate that bridges may be constructed to produce harmonized deadwood estimates based on reference definitions, regardless of the national definitions used to collect the data. Second, harmonizing estimates for categories of spatial position, decay class and species composition was relatively easy, although harmonization with respect to piece-and plot-level estimation was more difficult because expansive bridges were more frequently required. Third, as should be expected, harmonized estimates of deadwood volume based on different reference definitions may deviate considerably from estimates based on national definitions. However, rather large ranges of minimum diameter thresholds (10–20 cm) had little effects on the proportions of deadwood volume estimates by spatial position, decay class, species composition class, and forest category. Classification of lying deadwood and dead stems with respect to piecewise volume calculation methods (e.g., Huber, Smalian) would contribute to greater ease in harmonization. Fourth, acquisition of minimum and maximum diameters and the length between them for all cwd elements would greatly facilitate development of taper models and bridges. Thus, NFIs are encouraged to develop or adjust their field protocols so that comparisons among countries can be simplified via "harmonization ready" deadwood definitions, bridges, and field methods. Rondeux et al. (submitted) provide a detailed list of recommendations.

The aggregation of estimates from different countries that are based on different local deadwood definitions may lead to substantial decreases in the quality of those aggregated international estimates. Spatial variability and temporal changes in the amount of dead woody debris and in decay stages due to different silvicultural techniques, stand types, species composition, age, structure (Gore and Patterson 1986) could be confounded with differences due to local deadwood definitions adopted.

5.6 Ground Vegetation

5.6.1 Introduction

The analysis of the responses to the questionnaires (Sect. 3.7) suggests that indicators for the ground vegetation essential feature of forest biodiversity could be based on structural and compositional ground vegetation variables assessed by NFIs. Harmonization of estimates of the indicators appears feasible for three NFI variables: (1) ground vegetation layers, which are defined based on the different life-forms in different layer heights; (2) shrub cover, and (3) shrub species. Consequently, data in the COST Action E43 common database (Chap. 4) were used to investigate three indicators: (1) cover of ground vegetation layers expressed as a percentage of plot area, (2) presence/absence of shrub species, and (3) cover of shrubs expressed as a percentage of plot area. Harmonization results for these indicators are discussed in the following sections.

5.6.2 Materials

The common database includes raw data for 6,019 plots from 10 European countries stored in the common DB (Chap. 4). For harmonization of ground vegetation layers, data for 34,044 elements from 4,476 plots from eight countries were used; for harmonization of shrub species, data for 8,444 elements from 2,870 plots from six European countries were used; and for cover harmonization, data for 12,799 elements from 4,290 plots from six countries were used. However, the countries for which data were used differed for the three indicators.

- The harmonization test for the ground vegetation layers included data from the Walloon region of Belgium, the Czech Republic, Denmark, Germany, Italy, Portugal, Spain and Switzerland. Plots were distributed by country as follows: 100 plots (2%) from Italy, 250 plots (6%) from Portugal, 302 (7%) from Switzerland, 400 plots (9%) from Belgium, 401 plots (9%) from the Czech Republic, 775 plots (17%) from Spain, 790 plots (18%) from Germany, and 1,458 plots (33%) from Denmark.
- The harmonization test for shrub composition was based on data from Austria, Italy, Portugal, Spain, Sweden and Switzerland. Plots were distributed as follows: 188 plots (7%) from Sweden, 250 plots (9%) from Portugal, 251 plots (9%) from Italy, 302 plots (11%) from Switzerland, 775 plots (27%) from Spain, and 1,104 plots (38%) from Austria.
- The harmonization test for the shrub cover variable was based on data from Austria, Czech Republic, Denmark, Italy, Portugal, Spain and Switzerland. Plots were distributed as follows: 250 plots (6%) from Portugal, 302 plots (7%) from Switzerland, 401 plots (9%) from the Czech Republic, 775 plots (18%) from Spain, 1,104 plots (26%) plots from Austria and, 1,458 plots (34%) from Denmark.

To harmonize estimates using the proposed reference definitions, a full understanding of local methods and definitions adopted by the test countries was needed for the following ground vegetation topics: ground vegetation definitions, ground vegetation layers, species lists, sampling methods and cover scales used.

5.6.3 *Methods*

To harmonize estimates of indicators among countries, the first step is to construct common reference definitions, and the second step is to construct bridges that convert estimates based on national definitions to estimates based on the reference definitions (Vidal et al. 2008). For this study *reductive* and *neutral* bridges were used (Ståhl et al. submitted).

5.6.3.1 Ground Vegetation Layers

Our COST Action E43 reference definition for ground vegetation is: "Ground vegetation comprises all plants (excluding epiphytes) including tree species seedlings and saplings up to a height of 5 m". This is a broad concept for which identification of necessary field measurements and selection of subsequent NFI analyses is difficult. The next step, therefore, was to establish a *ground vegetation layer* reference definition using information on plant life-forms and height classes. However, because the country layer height ranges and life-forms used are quite different (Table 5.8), harmonization of classes is required. Nevertheless, our reference

Table 5.8 National description of ground vegetation layers and harmonization relative to the reference definition

Variable: Ground vegetation layer		
Country	National ground vegetation components	Harmonized layer relative to reference definition
Belgium (Walloon region)	Herb layer (all non ligneous and ligneous < 0.5 m) = Belgium herb layer (only non-ligneous including climbers)	Herb layer
	Ligneous layer <3 m = regeneration layer	Shrub layer
	Ligneous layer >3 m = "tree" layer	Not Harmonized
Switzerland	Herb layer (all non-ligneous, and ligneous < 0.5 m height)	Herb layer
Czech Republic	Grass	Herb layer
	Herb	Herb layer
	Fern	Herb layer
	Brush like herb	Not Harmonized
Germany	Bryophyta	Bryophyte layer
	Climbing plants	Shrub layer
	Dwarf shrubs	Not Harmonized
	Gramineae and Cyperaceae	Herb layer
	Half-shrubs	Not Harmonized
	Lichens	Bryophyte layer
	Pteridophyta	Herb layer
	Spermatophyta	Not Harmonized
	Shrubs 0.5–2 m high	Shrub layer
	Shrubs higher than 2 m	Shrub layer
	Shrubs less than 0.5 m high	Herb layer
Denmark	Bare soil	Not Harmonized
	Moss layer (terricolous bryophytes and lichens)	Bryophyte layer
	Terricolous lichens	Bryophyte layer
	Herb layer (all non-ligneous, and ligneous <0.5 m height)	Herb layer
Spain	Ferns	Herb layer
	Herbs	Herb layer
	Mosses, lichens and liverworts	Bryophyte layer
	Shrub	Shrub layer
Italy	Climbing plants	Shrub layer
	Herb	Herb layer
	Shrub	Shrub layer
Portugal	Herb layer: Regeneration/Plantation (trees)	Herb layer
	Herb layer: Herbs	Herb layer
	Shrubs/no perennial	Herb layer
	Shrub layer: Regeneration/Plantation (trees)	Shrub layer
	Shrub layer: Herbs	Shrub layer
	Shrub layer: Shrubs	Shrub layer
	Shrub layer: Climbers plants	Shrub layer

definition for ground vegetation layers follows from our analysis of the country definitions and includes the following components:

- The *shrub layer* includes ligneous plants with heights between 0.5 m and 5 m and may include regeneration (tree saplings due to the difficulty of estimating ground vegetation covers excluding them), shrubs and climbers; palm shrubs are also included.
- The *herb layer* includes all non-ligneous and ligneous herbs, ferns, shrubs and tree species with heights < 0.5 m. Bryophytes are excluded.
- The *bryophyte layer* includes all bryophytes and lichens growing on the ground.

Complete definitions would also require the definitions of the ground elements and the maximum heights of the vegetation elements included as ground vegetation. For example, it is common to record bryophytes on very small dead branches on the ground as ground vegetation, although bryophytes on big stones, rocks and large deadwood logs are not considered as ground vegetation. Details regarding the specifics of definitions for different NFIs are not available.

Harmonization of estimates of indicators for the shrub layer is difficult for two reasons. First, the incompleteness of inventories caused by restricted species lists and failure to assess all shrub layers is particularly problematic. Second, overlapping coverages of different shrubs cause problems when aggregating the coverages into a shrub cover sum. For example, Germany considers cover estimates for (1) dwarf shrubs, (2) sub-shrubs with 1–2 year-old suckers, (3) shrubs with heights less than 0.5 m, (4) shrubs with heights between 0.5 m and 2 m and (5) shrubs with heights greater than 2 m. If the total shrub cover is obtained by adding the covers of the sub-shrub layers, estimates may be greater than 100%, a result that is not comparable with results for countries whose cover values do not exceed 100%. Spain and Italy, however, assess cover by estimating entire shrub cover with a single estimate between 0 and 100% which enables direct comparisons of indicator estimates among the countries and for time series. The Walloon region of Belgium estimates coverages for four classes: (1) ligneous with heights less than 3 m, (2) ligneous with heights between 3 m and 10 m, and (3) ligneous with heights greater than 10 m, (4) herbaceous plants. In this case, the total shrub cover estimate could be based on species cover estimates because they monitor every species. However, the same problem occurs because the cover total could exceed 100%. Several approaches to a solution for summarising the sub-shrub layer covers are possible: (1) sub-shrub layer covers are summarized with a maximum of 100% or (2) formulas are used to estimate the degree of cover overlap which is then subtracted from the cover total. We propose the first option for which estimates over 100% are proportionally readjusted to 100% although errors due to overlapping are possible. This approach considers that the growing condition for shrub regeneration is best in the space between older shrubs in which case regeneration will mainly fill in the open spaces. This harmonization approach was used for comparison of Spanish pastures (San Miguel 2009). These options all merit further investigation.

Harmonization of estimates of indicators for the herb layer poses the additional challenge of establishing a harmonized record period because of the large temporal variation among country protocols. Additionally, numerous analyses have demonstrated an earlier onset of spring events for mid- and higher latitudes and a longer growing season because of climate change (Menzel et al. 2006) which is not considered in the NFI protocols. Monitoring dates in the common database were analysed, but the complexity necessary to harmonize phenological states for different altitude, latitude and others ecological factors is beyond the scope of this study. Perhaps harmonization of field recording dates must simply refer to the vegetation period, whenever this period starts or ends, under different climatic and topographical conditions with knowledge that particular species and cover changes occur during the vegetation period. However, variability among years is usually greater than variability within a vegetation period (White et al. 1997).

Because of the consistency among definitions, harmonization of estimates of indicators for the bryophyte layer seems feasible when the layer is monitored. For countries that disaggregate cover estimates for the bryophyte layer by species, the total cover can be obtained as a sum using the first approach described above. However, for countries that record cover in classes instead of percentages, obtaining the total as a sum is not accurate because mid-class values do not necessarily represent field cover. Also, changes in bryophyte cover are difficult to detect using mid-class values.

5.6.3.2 Shrub Species

Differences in species nomenclatures in the common database are a result of heterogeneous data sources. For harmonizing the species names, we used the European species International Cooperative Programme – Forests (ICP) codes (http://www.icp-forests.org/EPbiodiv.htm). These codes use three digits to identify the plant family, three digits for the genera and three digits for the species. For example, *Juniperus communis* is coded as 028.005.002 where the family is represented by 028 (Cupressaceae), the genus by 005 (Juniperus) and the species by 002. Thus, different taxonomical levels can be specified, recorded and analysed. To obtain comparable information for shrub species, harmonization must focus on two issues: (1) countries do not record the shrub species to the same level as the ICP codes, and (2) countries do not use complete lists of the shrub species that grow in their countries with the result that the absence of a species in the list does not document the absence of the species in the forest. Bridges for the shrub layer can be considered neutral for every country at all ICP levels by converting local or regional species, genera and family names into the ICP codes.

5.6.3.3 Cover Classes

Four countries (Austria, Denmark, Spain and Portugal) provide cover values in percentages from 0 to 100%; five countries (the Walloon region of Belgium, the

Czech Republic, Germany, Spain and Switzerland) use cover percentage classes; and one country, Italy, does not assess cover information. We propose two different reference definitions: (1) define cover percentages that represent the means of cover percentage classes, and (2) define a reference class cover scale.

Because a reductive bridge (Ståhl et al. submitted) is assumed to lead to harmonization more readily, we selected the second option (Table 5.9) for the test. With the exception of Germany, all countries already use the same boundaries for the higher percentage cover classes following the Braun-Blanquet scale (1965): 25–50%, 51–75% and 76–100%. However, because the majority of plant species naturally have small cover percentages, harmonization of classes with cover less than 25% is important. For the four countries using cover values (Austria, Denmark, Spain, and Portugal), we selected a reductive bridge; for Switzerland and Belgium neutral bridges were used. Cover classes for the Czech Republic were <1%; 1–9%; 10–25%; 26–50%; 51–75% and 76–100%, and for Germany the classes were 1–10%; 10–50% and >50%. For the Czech Republic, construction of a bridge was not possible because cover information for only one ground vegetation layer was available. In Germany no harmonization was possible because of the broad percentage classes. However harmonization could have been achieved by using the classes 1–5% and 5–25%. Further studies should be conducted to analyse distributions within forest types which requires more NFI data than represented in the common database.

Table 5.9 National ground vegetation cover scales and their harmonization (3rd column)

Variable: Cover scale		
Country	National cover scales	Reference cover scale
Austria	Value	Harmonized
Czech Republic	<1%	<1%
	1–9%	Bridge needed
	10–25%	[5–25]%
	26–50%	(25–50]%
	51–75%	(50–75]%
	75–100%	(75–100]%
Denmark	Value	Harmonized
Germany	1–10%	Not Harmonized
	10–50%	Not Harmonized
	>50%	Not Harmonized
Portugal	Value	Harmonized
Spain	Value	Harmonized
Switzerland	Up to 1%	<1%
	1–5%	[1–5]%
	6–25%	(5–25]%
	26–50%	(25–50]%
	51–75%	(50–75]%
	76–100%	(75–100]%

5.6.4 Results

5.6.4.1 Cover of the Ground Vegetation Layers

Harmonization of estimates for the indicator *ground vegetation layers* is restricted to reporting proportions of plots by different life-form layers by country (Fig. 5.25) and by forest category (Fig. 5.26) but not for detailed height layers (Sect. 5.6.2). The tables following Figs. 5.25 and 5.26 reveal that large percentages of plots provide ground vegetation information but that they do not conform to the harmonized life form categories, *bryophytes*, *herbs* and *shrubs*. Complete harmonization requires that life-form definitions and field methods are also harmonized as described in the previous section.

5.6.4.2 Shrub Harmonization

Investigations were conducted for shrub species and national cover scales with the objective of harmonizing estimates of two indicators: (1) presence/absence of shrub species, and (2) shrub species cover.

Data for the indicator tests consisted of national species. The indicator shrub species presence/absence could be considered as a substitute for cover assessment

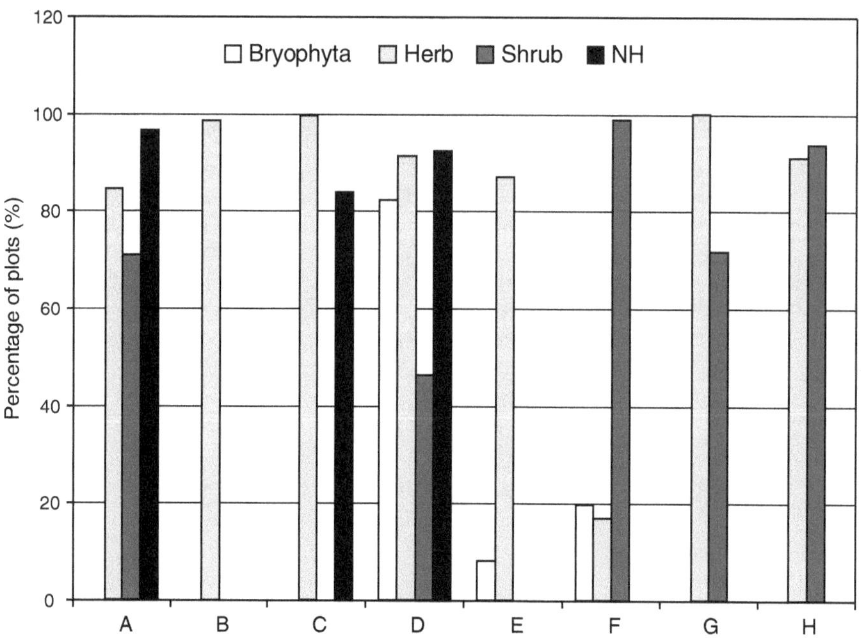

Fig. 5.25 Percentages of plots on which each vegetation layer life-form is observed by countries A–H (*NH* denotes plots whose estimates could not be harmonized)

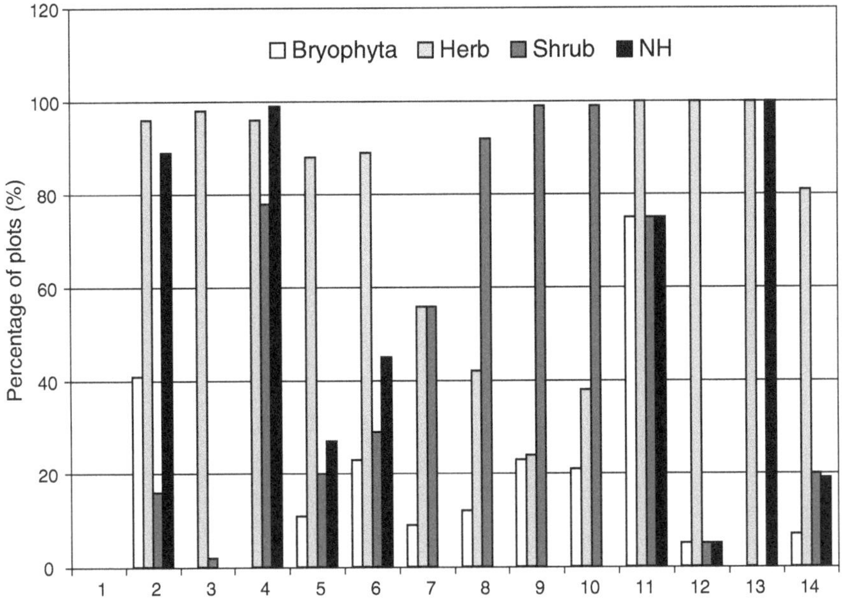

Fig. 5.26 Percentages of plots within the forest categories 1–14 on which each vegetation layer life-form is observed (*NH* denotes plots whose estimates could not be harmonized; see Table 5.2 for names of forest categories)

because of its simplicity and because it is less influenced by surveyor judgement than is cover assessment, although the sizes of the plots influence the results (Ståhl 2003). The common list obtained by aggregating national lists included 135 species with 82 genera. The number of plots with observations of a species was analysed as a percentage of the total number of plots, and the number of shrub species and genera observed were analysed as percentages of the total numbers of species and genera, respectively, belonging to the aggregated shrub species lists for all the countries. All three percentages were calculated for countries (Fig. 5.27) and for the 14 European forest categories (EEA 2006; Fig. 5.28).

5.6.4.3 Shrub Species and Genera Records

The study of species distributions for both countries and forest categories, yields interesting results. Species that are observed for a wide variety of ecological conditions by multiple countries as well as indicator species for the forest categories can be identified. Observations of the presence of shrub species may permit precise characterization of the EEA forest categories and monitoring of these shrub species over time. For example, *Hedera helix* is included in the species lists of four countries and in ten forest categories. Further, *Daphne mezereum* is observed in 11 forest categories, but it is included in the species lists of only three countries. *Vaccinium myrtillus* is observed in only two countries but in nine forest categories, whereas

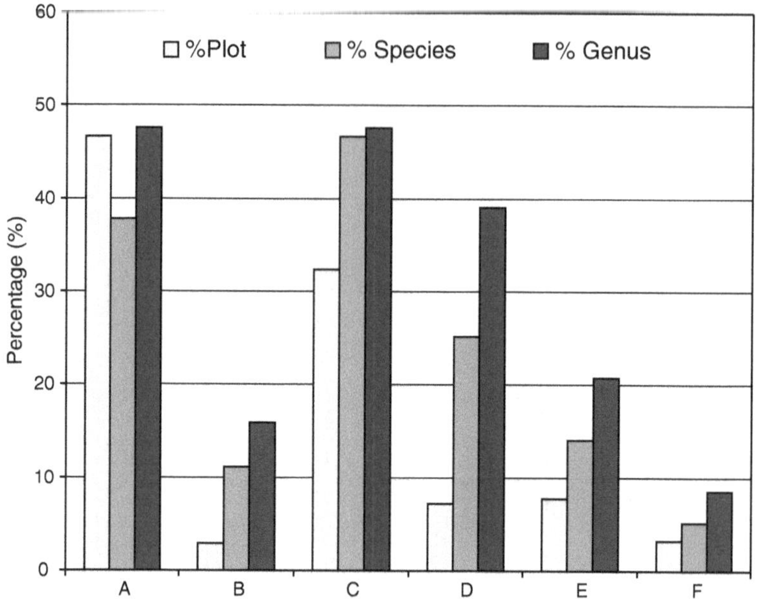

Fig. 5.27 Percentages of plots, percentages of observed species and percentages of observed genera for each country A–F

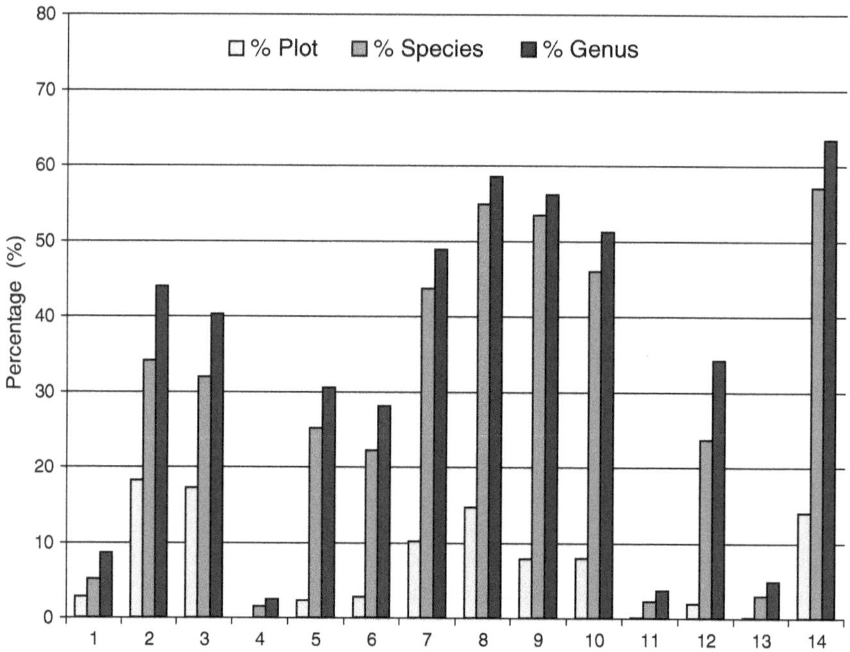

Fig. 5.28 Percentages of plots, percentages of observed species and percentages of observed genera by forest category 1–14 (see Table 5.2 for names of forest categories)

Cytisus sessilifolius appears only in one country but in four forest categories. *Rubus*, *Rosa* and *Daphne* are the three genera most frequently reported by NFIs.

5.6.4.4 Shrub Cover

This indicator adds additional information about species structure. The percentages of shrub species or genus observed for each reference cover scale class for each country are shown in Fig. 5.29 and for each forest category in Fig. 5.30. The large percentages in the lower classes of these figures are noteworthy but expected (*Cover classes*, Sect. 5.6.3) and justify their inclusion in the reference scales.

Only a few species naturally have large cover classes (>50%). For example, all observed species have covers less than 50% for forest categories 5, 6, 7, 9, 11 and 12, and approximately 80% of the observed species have covers less than 25% for the same forest categories. This result is consistent with the results of phytosociological surveys (Noest et al. 1989). *Vaccinium myrtillus* has large cover percentages in forest categories 2 and 3. In the Thermophilous deciduous forests (forest category 8), the poisonous shrub *Cytisus sessilifolius* is dominant with respect to cover; and in plantations (forest category 14) *Rubus* cover indicates high nitrogen supplies and light forest conditions.

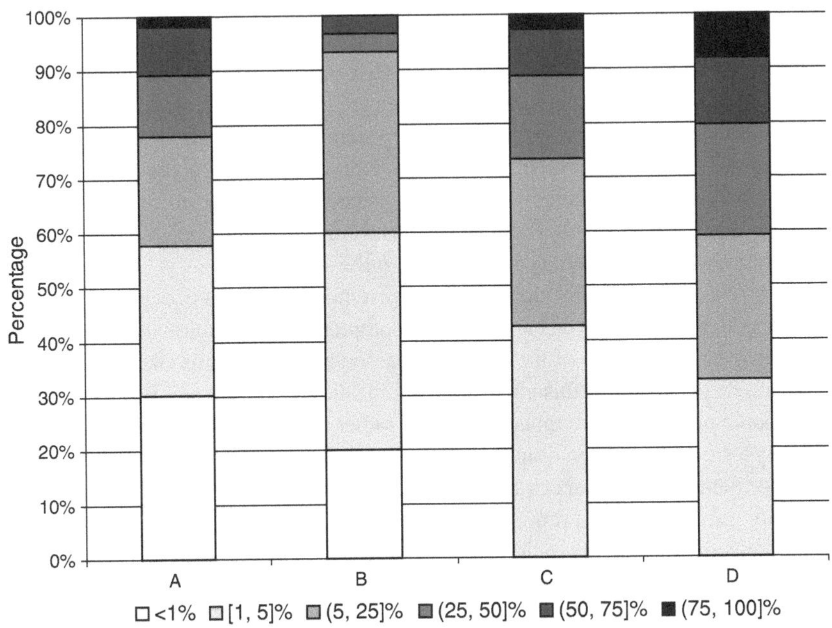

Fig. 5.29 Shrub cover category distributions by country A–D

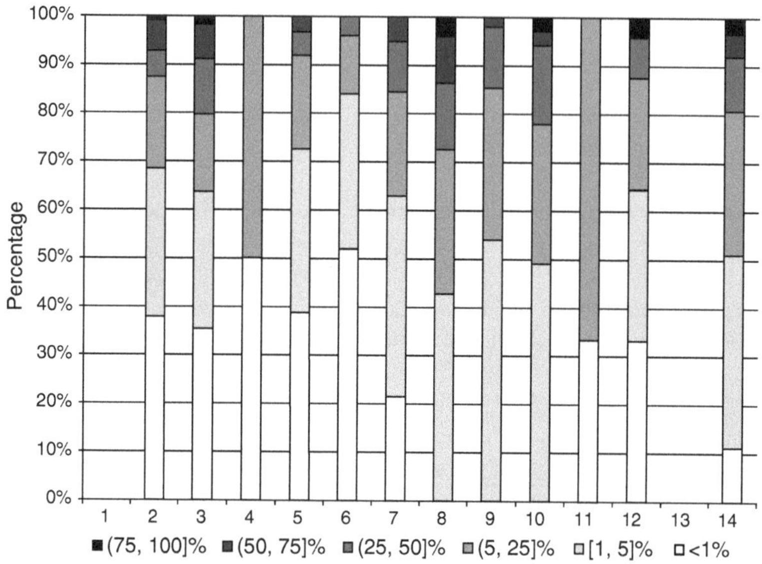

Fig. 5.30 Shrub cover category distributions by forest category 1–14 (see Table 5.2 for names of forest categories)

5.6.5 *Discussion*

Harmonization of estimates of ground vegetation layers is difficult because of different layer definitions used by countries, differences in monitoring seasons and overlapping cover and height classes for the life-forms. We propose a layer harmonization and approaches that aggregate cover values of sub-layers into main layers. For the indicator *shrub presence/absence*, the main challenge is that the species lists are not harmonized. Nevertheless, the observed shrub species that are part of each country's shrub species list may be reported using a common system of nomenclature such as the ICP codes. A harmonized shrub species list would substantially improve biodiversity monitoring in the future.

The sizes of monitoring areas and the inventory season are extremely relevant parameters that should also be harmonized if complete harmonization of ground vegetation is desired. As a result of the difficulties associated with harmonization of ground vegetation as described in this chapter, harmonization efforts would better be focused on harmonizing field measurement protocols rather than harmonizing estimates.

Analyses of information obtained from the questionnaires and from the common database identified differences among European countries with respect to ground vegetation monitoring and reporting. Recommendations to facilitate harmonization of estimates of ground vegetation indicators follow:

- Although complex classifications for describing ground vegetation biodiversity exist, a simple classification is needed to report on ground vegetation layers at the European scale.

- For purposes of ground vegetation reporting, the reference definition for the ground vegetation should specify all plants including tree species seedlings and saplings.
- Harmonized estimation of shrub composition indicators requires establishment of common inventory objectives (e.g. dominant species, endangered species, invasive species, indicator species) so that harmonized national shrub species lists can be constructed. Obviously, the greater the number of species on these lists, the better the biota can be described and the better the analysis can be conducted.
- To better harmonize estimates of cover indicators, cover values should be recorded in percentages from 0 to 100.

5.7 Naturalness

5.7.1 Introduction

The objective of the naturalness analyses was to investigate possibilities for harmonized estimation of indicators of forest naturalness using national forest inventory (NFI) data. Two complementary approaches to assessing naturalness were described in Sect. 3.8. The first approach is based on ecosystem processes and focuses on indicators of components of forest structure that characterize natural forests such as greater species diversity, greater horizontal diversity, and greater vertical diversity (McRoberts et al. submitted, McRoberts et al. 2009). The basic idea is that greater diversity indicates more natural forest. The second approach is based on the concept of hemeroby (Sukopp et al. 1990; Kowarik 1990) which is complementary to naturalness in the sense that it refers to the degree of human influence, rather than the lack of human influence. With this approach, visible signs of human influence such as cut stumps, plantings using regular spacings, and non-native species and less obvious signs such as decreased deadwood and increased volume all indicate less natural forest. Each approach has advantages and disadvantages.

The primary disadvantage of the hemeroby approach using NFI data is that signs of human influence as an NFI variable is dichotomous in the sense that observation of a single sign is all that is necessary to infer human influence. The consequence for assessments using NFI data is that the probability of observing such a sign is directly proportional to plot size. Thus, harmonized assessments require either common plot sizes among countries or knowledge of distances from plot centres to the locations of the signs of human influence. Because plot sizes vary widely among NFIs, and because locations on plots of signs of human influence are not recorded in general and, in particular, are not available in the common database, harmonized assessment of forest naturalness using the complementary hemeroby concept is nearly impossible. Therefore, the investigations were limited to the ecosystem processes approach that focuses on indicators of components of forest structure.

Forest naturalness studies using indicators of components of forest structure were investigated by McRoberts et al. (submitted) using NFI data from the United States of America (USA). These analyses focused on use of the Shannon index (Shannon 1949) as a measure of species richness and the standard deviation of diameter at-breast-height (dbh) as a measure of horizontal diversity. Measures of other components of forest structure such as vertical diversity, large or veteran trees, and age were not included because of their high correlations with measures of horizontal diversity based on dbh. Although deadwood was also acknowledged as an indicator of natural forest, deadwood data were not available for these cited studies.

The particular technical objective of the investigations reported in this chapter was to determine if the three indicators, deadwood volume per unit area, the Shannon index as a measure of species richness, and the standard deviation of dbh as a measure of horizontal diversity, all estimated in a harmonized manner using NFI data, could be used to identify plots with attributes assumed to be characteristic of natural forest. The underlying assumption was that more deadwood, greater species richness, and greater horizontal diversity are characteristics of more natural forest. Of crucial importance, however, is that the ecological validity of any approach to assessing natural forest cannot be readily or easily evaluated because of the near complete lack of natural forests in Europe. Thus, the emphasis was on evaluating the utility of the approach for harmonization purposes and the ecological reasonableness of the results. In addition, because the data in the common database were selected in different ways by different countries, and because the data do not represent random samples from individual countries, regions within countries, or even forest types or categories, the particular estimates obtained should not be construed as representative values.

5.7.2 Methods

The analyses required plot-level observations of deadwood and tree-level data in the form of species and dbh observations for individual trees. Harmonized plot-level estimates of deadwood volume per unit area, denoted *dwd*, as reported in Sect. 5.5 and by Rondeux et al. (submitted) were used. Because the Shannon index of species richness and the standard deviation of dbh were based on individual tree-level observations but estimated at plot level, and because plot sizes are not uniform among countries, the plot area on which each tree was measured was also required. Although some information on plot area was available in the common database, additional information was obtained from the country reports included in the COST Action E43 book (Tomppo et al. 2010). Finally, because the investigations focused on natural forest, data for plots from plantations and data for plots that included exotic species were excluded from all analyses. Based on these requirements, the natural forest investigations were based on data as summarized in Table 5.10.

Table 5.10 Numbers of plots by country and forest category

	Country					
Forest category[a]	Belgium[b]	Denmark	Italy	Spain	Sweden	Switzerland
1	0	0	0	0	142	0
2	0	0	0	0	0	0
3	0	0	0	0	0	201
4	95	0	0	0	0	0
5	92	334	0	0	0	0
6	89	113	0	0	0	100
7	0	0	18	123	0	100
8	0	0	90	232	0	0
9	0	0	48	124	0	0
10	0	0	62	107	0	0
11	0	0	0	0	0	0
12	0	0	10	0	0	0
13	0	19	0	0	6	0

[a] Table 5.2, EEA (2006)
[b] Walloon region of Belgium

Many countries use concentric circular plots for sampling trees of different sizes. For purposes of harmonized estimation, the varying sizes of these concentric plots for different countries and the associated varying minimum dbh thresholds must be accommodated if harmonization is to be realized. Thus, all plot-level estimates of the Shannon index and the standard deviation of dbh were scaled to a per unit area basis. The Shannon index, H', for species diversity was calculated as,

$$H' = -\sum_{i=1}^{s} p_i \cdot \ln(p_i) \tag{5.6}$$

where:

$$p_i = \frac{BA_{total,i}}{BA_{total}} \tag{5.7}$$

$BA_{total,\,i}$ is the total basal area for the ith species calculated as,

$$BA_{total,i} = \sum_{j=1}^{n_i} f_{ij} BA_{ij} \tag{5.8}$$

n_i is the number of trees of the ith species; $f_{ij} = \dfrac{10{,}000}{A_{ij}}$ is the number of trees per ha represented by the jth tree of the ith species; A_{ij} is the area of the plot in m^2 on which trees with dbh$_{ij}$ are observed; and BA_{ij} is the basal area of the jth tree of the ith species;

$$BA_{total} = \sum_{i=1}^{S} BA_{total,i} \tag{5.9}$$

and S is the total number of species observed on the plot. This estimator of H' assumes that all species observed on larger concentric circular sample plots are also observed on the smallest concentric sample plot of each main plot. This assumption will not always be valid with the result that H' may be biased downward. This estimator of H' also assumes that the distribution of trees by dbh is similar for concentric circular sample plots of all sizes associated with the same main plot.

The standard deviation of dbh was estimated as,

$$s_{dbh} = \sqrt{\frac{1}{\sum_{i=1}^{n} f_i} \sum_{i=1}^{n} f_i \cdot \left(dbh_i - \overline{dbh}\right)^2} \tag{5.10}$$

where $f_i = \frac{10,000}{A_i}$ is the number of trees per ha represented by the ith tree; A_i is the area of the plot on which trees with dbh_i are observed,

$$\overline{dbh} = \frac{1}{\sum_{i=1}^{n} f_i} \sum_{i=1}^{n} f_i \cdot dbh_i \tag{5.11}$$

and n is the number of trees observed on the plot.

A single index of naturalness was constructed that incorporated the contributions of all three indicators. First, the five largest plot-level values of dwd, H', and s_{dbh} over all plots without regard to country or forest category were determined; second, the means over these five largest values were calculated for each indicator; third, the plot-level estimates of dwd, H', and s_{dbh} were divided by their respective means to produce standardized values (Z_{dwd}", Z_H, Z_s) that, with only a few exceptions, were in the interval [0,1]. The exceptions were the standardized values for approximately half the five largest values which were slightly greater than one. For each plot, the single index, N', was calculated as,

$$N' = \sqrt{(Z_{dwd} - 1)^2 + (Z_H - 1)^2 + (Z_s - 1)^2} \tag{5.12}$$

Smaller values of N' indicate plots with all three values of dwd, H', and s_{dbh} closer to the means over the five maximum values; i.e., plots that are expected to be more similar to plots with a greater degree of naturalness. The distributions of N' were investigated for $dbh_{min} = 0$ cm and $dbh_{min} = 10$ cm for all plots for the six countries identified in Table 5.10. Estimates of indicators of forest structural diversity are known to be sensitive to plot size (McRoberts et al. 2009, submitted), but the effects could not be assessed for this study because distances from plot centers to individual trees were not included in the common database. Therefore, one set of analyses was restricted to data for Denmark, Italy, Sweden and Switzerland whose maximum plot sizes were similar, 500–707 m^2 (Table 5.11).

Table 5.11 Effects of minimum dbh over all forest categories (minimum 5 trees/plot)

Country	Maximum plot area (m^2)	Minimum dbh (cm)	No. plots	No. trees	Mean per plot Dwd (m^3/ha)	H′	S_{dbh} (cm)
A	1,017	0.0	276	2,865	6.45	0.54	11.37
		5.0	276	2,865	6.45	0.54	11.37
		10.0	276	2,468	6.35	0.49	10.45
B	707	0.0	466	8,773	2.38	0.39	6.28
		5.0	454	7,921	2.44	0.38	5.91
		10.0	435	6,698	2.53	0.37	5.57
C	531	0.0	231	1,600	3.50	0.31	7.28
		5.0	231	1,600	3.50	0.31	7.28
		10.0	230	1,478	3.43	0.28	6.65
D	1,963	0.0	586	13,476	1.65	0.27	9.36
		5.0	586	13,476	1.65	0.27	9.36
		10.0	586	12,506	1.65	0.27	9.12
E	500	0.0	148	1,353	5.39	0.40	3.71
		5.0	148	1,353	5.39	0.40	3.71
		10.0	148	1,353	5.39	0.40	3.71
F	500	0.0	401	4,240	10.78	0.56	9.94
		5.0	401	4,240	10.78	0.56	9.94
		10.0	401	4,240	10.78	0.56	9.94

5.7.3 *Results*

The effects of dbh_{min} on mean *dwd*, mean s_{dbh}, and mean H' were minimal (Table 5.11), although greater dbh_{min} indicated slightly smaller values for mean s_{dbh}, and mean H'. This result is consistent with the results reported by (McRoberts et al. 2009, submitted) using NFI data for the USA. For countries E and F, no changes in means with respect to dbh_{min} occurred because $dbh_{min} = 10$ cm for observations in the common database. Finally, greater values of dbh_{min} suggested slightly greater values for mean *dwd*.

Correlations among plot level values of *dwd* and s_{dbh}, and among plot-level values of *dwd* and H' were less than 0.005 and non-significant ($\alpha = 0.05$); the correlation among plot-level values of s_{dbh}, and H' was statistically significant ($\alpha = 0.05$) but was small, $\rho < 0.24$. These small correlations indicate that each indicator represents an independent dimension or axis for assessing naturalness. Alternatively, the three indicators represent different aspects of biodiversity and, as such, complement each other and contribute to a more complete assessment of forest naturalness.

The histograms of N′ indicated few small values as would be expected for Europe where there is little natural forest (Fig. 5.31). When analyses were based on data for the six countries, the ten plots with the smallest values of N′ were nearly the same for $dbh_{min} = 0$ and $dbh_{min} = 10$. These plots tended to come from Denmark, Sweden and Switzerland. Forest category, number of species, number of trees, and mean dbh for these plots varied considerably, although all ten plots tended to have at least one large tree. When analyses were restricted to data for

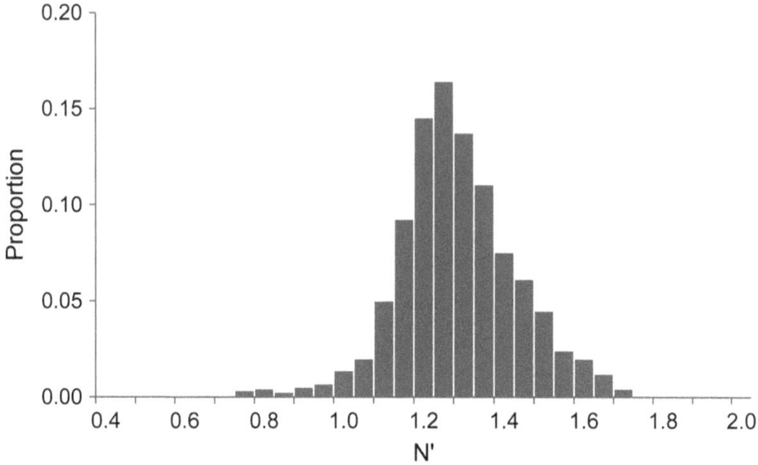

Fig. 5.31 Distribution of N′ for Denmark, Italy, Sweden and Switzerland whose maximum NFI plot sizes were similar

Table 5.12 Plots with least index, N′[a]

Rank	N′	Dwd (m^3/ha)	s_{dbh} (cm)	H′	Country	Foresttype	No. species	No. trees	Mean dbh (cm)	Max dbh (cm)
1	0.65	176.66	17.32	0.99	F	7.2	3	14	38.62	79.00
2	0.75	93.03	15.50	1.35	E	1.1	8	32	11.78	67.00
3	0.77	121.93	13.30	1.12	E	1.1	4	34	8.93	68.00
4	0.78	110.28	9.97	1.63	E	1.1	7	45	10.34	66.40
5	0.79	91.43	11.18	1.75	E	1.1	9	45	8.30	41.00
6	0.80	70.44	15.40	1.56	C	8.8	8	25	6.76	62.50
7	0.81	52.05	20.25	1.52	B	6.1	6	32	11.73	75.30
8	0.81	65.02	17.11	1.44	C	10.2	5	35	7.59	66.00
9	0.82	138.25	22.29	0.51	F	3.2	3	7	25.46	89.00
10	0.85	116.98	12.45	1.04	F	7.2	4	13	43.28	59.00

[a] Based on data for Denmark, Italy, Sweden and Switzerland whose maximum plot sizes were similar; dbh_{min} = 10 cm

Denmark, Italy, Sweden and Switzerland whose maximum plot sizes were similar, the ten plots with the smallest values of N′ included at least one from each country (Table 5.12). Again, forest category, number of species, number of trees, and mean dbh varied considerably, and all ten plots tended to have at least one large tree.

5.7.4 Discussion

Harmonized assessments of natural forest require initial harmonization of the components of indicators of natural forest; for this study the indicators were deadwood volume per unit hectare, the Shannon index (H') as an indicator of species composition, and the standard deviation of dbh (s_{dbh}) as an indicator of horizontal forest structural diversity. Harmonization of deadwood estimates is a difficult task and is documented elsewhere (Chap. 5.5, Rondeux et al. submitted). Harmonization of H' and s_{dbh} among countries requires accommodation for differences in minimum dbh thresholds and plot sizes. Although standardization of minimum dbh thresholds and plot sizes among countries would be advantageous for harmonization purposes, individual countries have selected these features of their NFIs to accommodate their unique ecological, climatic, topographic and economic situations and cannot be expected to change them. Thus, bridges to harmonize estimates of indicators using data collected with different NFI features are necessary (Ståhl et al. submitted). The bridge for harmonization with respect to minimum dbh threshold was reductive in that all trees with dbh < 10 cm were excluded from the analyses because for two countries, Sweden and Switzerland, all trees represented in the database had dbh ≥ 10 cm. Harmonization with respect to plot size was not possible because a bridge would have required distances between plot centers and tree locations which were not available. Therefore, the primary analyses used only data for four countries whose maximum plot sizes were similar, 500–707 m^2.

A single index, N′, was defined in terms of standardized plot-level values of the three indicators and was estimated using only NFI data. The index produced reasonable results under the assumption that more natural forest is characterized by greater volumes of deadwood per unit area, greater species richness, and greater dbh diversity. In particular, the index identified only a few plots in the left tail of the distribution of N′ with small values. However, the ecological validity of the index could not be evaluated because of the lack of information for forest areas known or assumed to be natural or near natural.

Additional analyses should focus in three areas. First, sufficient data should be acquired to permit analyses within each forest category. Second, data for probability-based samples for large areas should be acquired to permit valid estimates of population parameters. Third, the utility of the index, N′, should be evaluated more thoroughly to determine if it correctly identifies plots that are assessed as more natural using other approaches such as those based on the hemeroby concept or indicators with a more ecological basis (Winter et al. 2010).

5.8 Conclusions

Chapter 5 reports the results of the harmonization tests for indicators of forest biodiversity grouped into seven essential features (Table 2.6). The chapter authors tested 20 indicators using data from the common database described in Chap. 4. The methods varied considerably and are difficult to generalize or summarize.

Nevertheless, the general approaches used for the tests can be characterized as having four steps:

1. The forest biodiversity indicator must be defined based on international reference definitions for which agreement has been reached; in this step, the NFI variables necessary for estimation of the indicator must be selected;
2. For each NFI variable, an international reference definition must be defined;
3. NFI raw data must be processed to produce values of NFI variables;
4. Values of NFI variables are used to estimate harmonized forest biodiversity indicators.

A few concluding comments based on the results reported for the essential features are appropriate. The European Forest Types (EFT) system of nomenclature developed by the EEA (2006) was useful for comparing estimates of forest biodiversity indicators after aggregating data from the NFIs (Sect. 5.2). The procedure for the classification of an NFI plot based on the EFT system is frequently nearly automatic and thus very objective. In most cases, the only NFI raw information necessary is the geographic location of the plot and tree species and dbh. Currently, the EFT system is oriented toward implementation in the field. For this reason, the classification of some plots using NFI data produced less objective results. We strongly recommend standardized NFI procedures for acquisition of forest type information, in particular that the information be acquired in the field.

For forest structure (Sect. 5.3), a large number of indicators may be estimated using data for a relatively small number of NFI variables. The indicators for forest structural diversity focus primarily on evaluating plot-level horizontal, vertical and compositional tree diversity. The information needed for the estimation of these biodiversity indicators is just tree species, dbh, and height. Future effort should emphasise harmonization of vertical tree layer classes and evaluation of the nativeness of tree species.

Forest age (Sect. 5.4) is a crucial information item because the potential biodiversity value of a forest is closely related to age. The concept of old growth forest in such a sense is strictly related to the age of the trees. Although the definition of tree age is standardized at the international level, such is not the case for forest age. Thus, the indicators proposed for the harmonization test are quite innovative. Tree age is relative in the sense that different tree species have different natural life spans, and natural life spans are related to environmental conditions. The authors proposed a draft table of natural life spans for several tree species in different environmental conditions.

Simple and ecologically meaningful forest age indicators may be estimated using tree age data. Unfortunately the ages of all plot trees are not frequently recorded by NFIs. Although tree dbh can be used as a simple proxy variable, acquisition of tree ages, at least for the dominant, largest, and/or oldest trees is recommended. The authors' results suggest that the best indicator of forest age is dominant age defined as the mean age of the dominant trees in a stand or its proxy, dominant dbh defined as the mean dbh of dominant trees in a stand.

Deadwood (Sect. 5.5) is a classical biodiversity indicator because it has been adopted by many forest monitoring programmes. The basic indicator tested was the total volume of deadwood per hectare classified by broad tree species groups, spatial position and decay stage. The test focused on evaluating the impacts of different deadwood definitions on total volume estimates. The results demonstrated that the harmonization of deadwood volume and decay stage is feasible, even if the necessary bridges are often complex. Harmonization by deadwood species is not currently feasible, so the focus is on just differentiating coniferous/broadleaves groups. Finally the harmonization of spatial position (lying/standing) of deadwood is also not currently feasible because most countries classify the deadwood element using different local definitions without recording any quantitative data. We recommend that NFIs adopt local definitions for deadwood that are more amenable to harmonization via simple bridges.

Results for ground vegetation (Sect. 5.6) indicate that NFI information for this essential feature is still relatively sparse and that definitions and methods related to data acquisition differ considerably. Thus, the harmonization test was conducted using simple indicators which probably have limited value for the evaluation of the overall level of forest biodiversity. Nevertheless, these first experiences contribute to understanding the kind of field data that should be acquired for a more complete assessment of forest biodiversity that is not limited just to tree components. NFI programmes interested in more complete forest biodiversity assessments should invest in this effort.

In Sect. 5.7, a completely new indicator of forest naturalness was proposed. Because different authors have reported different interpretations of forest naturalness, no consensus has been reached regarding a unique definition of the term. Because of the complexity of this concept, the proposed indicator is also complex in the sense that is based on three different sub-indicators: deadwood volume; tree species diversity quantified using the Shannon index of basal area of tree species; and horizontal diversity quantified using the standard deviation of tree dbh. Even if the number and particular sub-indicators used change, the general approach and the manner in which the data are aggregated to derive a complex naturalness index have future potential for operational applications.

We acknowledge that circumventing the effects of plot size on harmonized estimation of indicators is still an open problem. Many of the indicators tested are sensitive to plot area, while the effects for other indicators are yet unknown (McRoberts et al. 2009, McRoberts et al. submitted). Although this problem could be completely resolved by standardising plot configurations and field methods, we recognize that plot configurations for individual NFIs have evolved over time to accommodate unique climatic, topographic, commercial and ecological factors. For indicators that require only tree data, considerably progress toward harmonization could be achieved if NFIs would record the geographic location on the plot of each tree. Despite difficulties associated with different plot sizes, harmonization tests evaluated several indicators and reported results for NFIs with plots of similar sizes.

References

Bailey, R. G. (1976). *Ecoregions of the United States (map)*. Ogden: USDA Forest Service, Intermountain Region.

Bailey, R. G. (1983). Delineation of ecosystem regions. *Environmental Management, 7*, 365–373.

Barbati, A., & Marchetti, M. (2004). Forest Types for Biodiversity Assessment (FTBAs) in Europe: the revised classification scheme. In M. Marchetti (Ed.), Monitoring and indicators of forest biodiversity in Europe – from ideas to operationality. EFI Proceedings No. 51. pp. 105–126.

Barbati, A., Corona, P., & Marchetti, M. (2007). A forest typology for monitoring sustainable forest management: the case of European forest types. *Plant Biosystems, 141*(1), 93–103.

BirdLife international. (2000). *Threatened birds of the world*. Barcelona and Cambridge: Lynx Ediciones and BirdLife International.

Bohn, U., Gollub, G., & Hettwer, C. (2000). *Map of the natural vegetation of Europe*. Bonn: Federal Agency for Nature Conservation.

Bradshaw, R. H. W., & Møller, P. F. (2004). European forest types for biodiversity assessment. A qualitative approach. In M. Marchetti (Ed.), Monitoring and indicators of forest biodiversity in Europe – from ideas to operationality. EFI Proceedings No. 51. pp. 127–133.

Braun-Blanquet, J. (1965). *Plant sociology; the study of plant communities*. London: Halner.

Corona, P., & Ferrara, A. (1992). Un metodo semplificato per la trasformazione delle tavole a doppia entrata per la cubatura di fusti interi in tavole assortimentali. *L'Italia Forestale e Montana, 3*, 144–157.

Davis, C. E., Moss, D., & Hill, M. (2004). EUNIS habitat classification. Revised 2004. Report to the European Environmental Agency. Available on-line at: eunis.eea.europa.eu/upload/EUNIS_2004_report.pdf. Accessed December 2010.

EEA (2006). European forest types. Categories and types for sustainable forest management and reporting. EEA Technical report No. 9/2006.

Fischer, R., Granke, O., Chirici, G., Meyer, P., Seidling, W., Stofer, S., Corona, P., Marchetti, M., & Travaglini, D. (2009). Background, main results and conclusions from a test phase for biodiversity assessments on intensive forest monitoring plots in Europe. *iForest, 2*, 67–74.

Gore, J. A., & Patersson, W. A. (1986). Mass of downed wood in northern hardwood forests in New Hampshire: potential effects of forest management. *Canadian Journal of Forest Research, 16*, 335–339.

Helms, J. A. (Ed.). (1998). *The dictionary of forestry* (p. 210). Bethesda: Society of American Foresters. ISBN 0-939970-73-2.

ICP – International Co-operative programme on assessment and monitoring of air pollution effects on forests. Plant code lists: http://www.icp-forests.org/DocsBiodiv/Vol1.TXT. Accessed July 2009.

Kändler, G., & Riemer, U. (2005). Der Zusammenhang zwischen Wuchsleistung und Standort bei Fichte, Tanne und Buche auf den wichtigsten Standortseinheiten des Südschwarzwaldes. In K. von Teuffel, M. Baumgarten, M. Hanewinkel, W. Konold, U. H. Sauter, H. Spiecker, & K. von Wilpert (Eds.), *Waldumbau für eine zukunftsorientierte Waldwirtschaft* (p. 422). Berlin, Heidelberg: Springer.

Kowarik, I. (1990). Natürlichkeit, Naturnähe und Hemerobie als Bewertungskriterien. In H. Sukopp, S. Hejný, & I. Kowarik (Eds.), *Urban ecology*. The Hague: SBP Academic Publications.

Larsson, T.-B., Svensson, L., Angelstam, P., Balent, G., Barbati, A., Bijlsma, J. R., Boncina, A., Brad-shaw, R., Bücking, W., Ciancio, O., Corona, P., Diaci, J., Dias, S., Ellenberg, H., Fernandes, F. M., Fernàndez-Gonzalez, F., Ferris, R., Frank, G., Møller, PF., Giller, P. S., Gustafsson, L., Halbritter, K., Hall, S., Hansson, L., Innes, J., Jactel, H., Keannel Doppertin, M., Klein, M., Marchetti, M., Mohren, F., Niemelä, P., O'Halloran, J., Rametsteiner, E., Rego, F., Scheidegger, C., Scotti, R., Sjöberg, K., Spanos, I., Spanos, K., Standovár, T., Tømmerås, Å., Trakolis, D., Uuttera, J., Walsh, P. M., Vande-kerkhove, K., Watt, A. D, & VenDenMeersschaut, D. (2001). Biodiversity evaluation tools for European forests. A report from the FAIR project

"Indicators for monitoring and evaluation of forest biodiversity in Europe" CT97-3575 within the EU Commission RTD Programme. *Ecological Bulletin, 50*, 150.

Lindner, M., Maroschek, M., Netherer, S., Kremer, A., Barbati, A., Garcia-Gonzalo, J., Seidl, R., Delzon, S., Corona. P., Kolström, M., Lexer, M. J., & Marchetti, M. (2009). Climate change impacts, adaptive capacity, and vulnerability of European forest ecosystems. *Forest Ecology and Management.* doi:10.1016/j.foreco.2009.09.023.

McRoberts, R. E., Winter, S., Chirici, G., & LaPoint, E. (submitted). Assessing forest naturalness. *Forest Science.*

McRoberts, R. E., Tomppo, E. O., Schadauer, K., Vidal, C., Ståhl, G., Chirici, G., & Smith, W. B. (2009). Harmonizing national forest inventories. *Journal of Forestry, 107*, 179–187.

McRoberts, R. E., Winter, S., Chirici, G., Hauk, E., Pelz, D. R., Moser, W. K., & Hatfield, M. A. (2008). Large-scale spatial patterns of forest structural diversity. *Canadian Journal of Forest Research, 38*, 429–438.

Menzel, A., Sparks, T. H., Estrella, N., Koch, E., Aasa, A., Ahas, R., Alm-Kübler, K., Bissolli, P., Braslavská, O., Briede, A., Chmielewski, F. M., Crepinsek, Z., Curnel, Y., Dahl, Å., Defila, C., Donnelly, A., Filella, Y., Jatczak, K., Måge, F., Mestre, A., Nordli, Ø., Peñuelas, J., Pirinen, P., Remisová, V., Scheifinger, H., Striz, M., Susnik, A., Wielgolaski, F. -E., v Vliet, A., Zach, S., & Zust, A. (2006). European phenological response to climate change matches the warming pattern. *Global Change Biology, 12*, 1969–1976.

Noest, V., van der Maarel, E., van der Meulen, F., & van der Laan, D. (1989). Optimum-transformation of plant-species cover-abundance values. *Vegetatio, 83*(1–2), 167–178.

Pardé, J. (1956). Une notion pleine d'intérêt: La hauteur dominante des peuplements forestiers. *Revue Forestière Française, 12*, 850–856.

Pearman, P., Penskar, M. R., Schools, E. H., & Enander, H. D. (2006). Identifying potential indicators of conservation value using natural heritage occurrence data. *Ecological Applications, 16* (1), 186–201.

Rodwell, J., Schaminée, J., Mucina, L., Pignatti, S., Dring, J., & Moss, D. (2002). The diversity of European vegetation. An overview of phytosociological alliances and their relationships to EUNIS habitat. Wageningen: Naturbeheer envisserij/European Environment Agency.

Rondeux, J., Bertini, R., Bastrup, A., Corona, P., McRoberts, RE., Sanchez, C., Ståhl, G., Winter, S., & Chirici, G. (submitted). Assessing deadwood using harmonised national forest inventory data. *Forest Science.*

Rondeux, J., Bertini, R., Bastrup-Birk, A., Corona, P., McRoberts, R. E., Sanchez, C., Ståhl, G., Winter, S., & Chirici, G. (submitted). Harmonised deadwood assessments using national forest inventory data. *Forest Science.*

San Miguel, A. (Coord.). (2009). Los Pastos de la Comunidad de Madrid. Tipología, Cartografía y Evaluación. Serie Técnica de Medio Ambiente no 4. Comunidad de Madrid. Madrid.

Shannon, C. E. (1949). The mathematical theory of communication. In C. E. Shannon & W. Weaver (Eds.), *The mathematical theory of communication*. Urbana: University of Illinois Press.

Ståhl, G. (2003). Presence/absence sampling as a substitute for cover assessment in vegetation monitoring. In P. Corona et al. (Eds.), *Advances in forest inventory for sustainable forest management and biodiversity monitoring* (pp. 137–142). Dordrecht: Kluwer Academic Publishers.

Ståhl, G., Cienciala, E., Chirici, G., Gabler, K., Lans, A., Vidal, C., Winter, S., McRoberts, R. E., Schadauer, K., & Tomppo, E. (submitted). Bridging national and reference definitions for harmonised provision of forest statistics. *Forest Science.*

Sukopp, H., Hejný, S., & Kovarik, I. (Eds.). (1990). *Urban ecology: plans and plan communities in urban environments*. The Hague: SPB Academic Publications.

Tomppo, E., Gschwanter, T., Lawrence, M., & McRoberts, R. E. (2010). *National forest inventories: pathways for common reporting* (671 pp). Heidelberg: Springer.

U.S. Forest Service (2007). Forest inventory and analysis national core field guide, Vol. 1: Field data collection procedures for phase 2 plots, version 4.0. Northern Research Station, Forest Service, U.S. Department of Agriculture. (292 pp.).

Vidal, C., Lanz, A., Tomppo, E., Schadauer, K., Gschwantner, T., Di Cosmo, L., & Robert, N. (2008). Establishing forest inventory reference definitions for forest and growing stock: a study towards common reporting. *Silva Fennica, 42*(2), 247–266.
White, M. A., Thornton, P. E., & Running, S. W. (1997). A continental phenology model for monitoring vegetation responses to inter-annual climatic variability. *Global Biogeochemical Cycles, 11*, 217–234.
Whittaker, R. H. (1972). Evolution and measurement of species diversity. *Taxon, 21*, 213–251.
Winter, S., Fischer, H. S., & Fischer, A. (2010). Relative quantitative reference approach on naturalness assessments. *Forest Ecology and Management, 259*, 1624–1632.

Chapter 6
Summary and Conclusions

Gherardo Chirici, Susanne Winter, and Ronald E. McRoberts

Abstract This chapter summarizes the main findings of WG3 of COST Action E43 by recalling the background information which constitutes the reasons for the activities carried out and then recalling the main phases of our work. For each of the seven essential features, the main results of the experimental harmonization process are presented and recommendations to NFIs are given to facilitate future operational harmonization processes.

6.1 Introduction

Historically, national forest inventories (NFI) were developed for purposes of estimating the economic value of forests, mainly from timber and wood production. However, the value of forests is no longer strictly related to their economic value. Although a forest is considered an economic resource because of its capacity to produce wood and non-wood products, it is not considered only as wood factory. Forests have intrinsic environmental, social and aesthetic values. NFIs have gradually accepted this new holistic vision of forest resources and have modified their field protocols to include new variables useful for assessing the overall set of goods and services provided by forest habitats: from CO_2 sequestration to protection

G. Chirici (✉)
Università degli Studi del Molise, Italy
e-mail: gherardo.chirici@unimol.it

S. Winter
Department für Ökologie und Studienfakultät für Forstwissenschaft und Ressourcenmanagement, Technische Universität München, Germany
e-mail: winter@wzw.tum.de

R.E. McRoberts
Forest Inventory and Analysis, Northern Research Station, USDA Forest Service, USA
e-mail: rmcroberts@fs.fed.us

G. Chirici et al. (eds.), *National Forest Inventories: Contributions to Forest Biodiversity Assessments*, Managing Forest Ecosystems 20,
DOI 10.1007/978-94-007-0482-4_6, © Springer Science+Business Media B.V. 2011

against floods and avalanches; from landscape to social and recreational values; and from the production of non-wood goods to the protection of habitats and biodiversity. This new generation of NFIs is frequently characterized as multipurpose.

This publication is the result of the activities of Working Group 3 (WG3) of COST Action E43 between 2004 and 2009. The activities focused on evaluating the potential role of NFIs as a source of information for harmonized international reporting on forest biodiversity. Harmonization is essential for valid comparison of trends in indicators across countries.

Our approach to evaluating the value and usefulness of NFIs as a source of information for forest biodiversity monitoring was data driven. In fact, the term biodiversity has a very wide meaning including different scales of observation. Because of economic constraints, it is not possible to monitor forest biodiversity from the perspective of its most common meaning "the diversity of life in all its forms and all its levels of organization within forested areas" (Hunter 1990). Therefore, to monitor biodiversity it is essential to express this very wide meaning in a smaller set of indicators specifically designed for monitoring purposes.

The selection and definition of biodiversity indicators is a complex process. The indicators should be proxy variables for "biodiversity," but because "biodiversity" was never measured, its definition cannot be based on an experimental approach. Experts are frequently requested to address this conceptual problem. An indicator is, therefore, developed when a large number of experts come to common consensus on its relevance (Stokland et al. 2003). Such an approach has already been used by multiple international processes to produce sets of biodiversity indicators for international reporting (EEA 2007; 2009; CBD 2007; MCPFE 2003; Montréal Process 2006; FAO 2005). The process used by WG3 of COST E43 was a similar phased approach with consequent milestones. Here we recall the main milestones.

In the early phase of the project, the activities of WG3 were divided into working subgroups. One subgroup was established for each of the essential features of forest biodiversity (Table 2.6). The essential features were selected on the basis of responses to a questionnaire compiled by NFI and forest biodiversity experts. The questionnaire was used to rank a large set of possible forest variables on the basis of their importance for monitoring forest biodiversity and on their potential feasibility for implementation in standard NFI field protocols (Winter et al. 2008). Seven essential features were identified: forest categories, forest age, forest structure, deadwood, regeneration, ground vegetation, and naturalness.

In the second phase, local definitions, sampling methods and estimation procedures adopted by the NFIs were described for NFI variables associated with essential features. After analyzing differences and similarities among countries, WG3 decided to include an experimental phase aimed at testing possible bridges for converting estimates based on local definitions to estimates based on international reference definitions (Ståhl et al. submitted).

For this purpose, the third phase of the project consisted of constructing a common database that was populated with data for 14,638 forest inventory plots from 13 European countries and the USA. Each country used its own criteria for selecting plots to be included in the common database which means that the data cannot be considered a representative sample for all the considered countries. However, the nature of the data has no adverse effect because the objective was to evaluate the feasibility of constructing bridges, not to operationally implement the harmonization process itself. Thus, the test results should be considered as a methodological contribution for possible future application at country level. Further, inferences based on the test results cannot be extended beyond the common database.

The findings of WG3 of COST Action E43 were that the historical "wood oriented" perspective is still common among NFIs and, consequently, most information acquired in the field is still related to trees. Further, although a complete set of harmonized biodiversity indicators may be estimated for the tree component of a forest, few or no harmonized indicators may be currently estimated for other populations of forest flora and fauna. Thus, because the tree component of forests is still the component of primary human interest, the NFI vision of forest biodiversity must still be considered anthropocentric. Although trees may be the most important forest component from a limited human perspective, from a wider biological perspective the populations of all life-forms interacting with a forest habitat potentially have the same ecological relevance. Thus, all populations and components should theoretically be assessed in a complete biodiversity monitoring programme.

The list of indicators proposed by WG3 was limited by the availability of NFI data used to populate the common DB. These data represent the current contributions of NFIs to forest biodiversity monitoring within the framework of COST Action E43, not the potential best set of indicators for monitoring forest biodiversity nor the set required to completely satisfy international reporting commitments.

6.2 Overview of the Results from the Harmonization Tests

The approach followed in the harmonization tests consists of two steps: (*i*) harmonized forest biodiversity indicators were estimated at plot level, and then (*ii*) plot-level data and/or estimates were aggregated and reported for geographical regions (countries) or by forest categories (EEA 2006) (Table 5.2). The complete list of indicators reported in the sections of Chap. 5 is shown in Table 6.1.

To harmonize estimates of indicators, the underlying NFI variables on which the indicators are based must also be harmonized with respect to aspects such as sampling protocols and measurement thresholds in accordance with international references. In WG3 we applied the general reference definitions jointly developed by all the working groups of COST Action E43 and some others specifically proposed for forest biodiversity (Tomppo et al. 2010).

Table 6.1 List of forest biodiversity indicators considered in the harmonization tests presented in Chap. 5. For each essential feature, plot-level indicators are presented (second column); for each plot-level indicator, the necessary raw NFI data are listed in the third column; and finally the aggregated indicators are listed in the last column. The indicator 2.8 (Shannon index of tree species) was proposed as one of the component of index 4.1 (Naturalness)

Code	Indicator	Input data to be harmonized/ standardized	Aggregation per forest categories or geographical areas
1.1	Forest category (EEA 2006)	1 – basal area by tree species	
		2 – geographic position	
2.1	Relative abundance of native tree species in terms of basal area	1 – nativeness of tree species 2 – basal area by tree species	Relative abundance of native tree species in terms of basal area
2.2	Number of native tree species	1 – nativeness of tree species	Mean number of native tree species
2.3	Proportion of plots with 1, 2, 3 and more native tree species	1 – nativeness of tree species	Proportion of plots with 1, 2, 3 and more native tree species
2.4	Largest diameter trees	1 – tree dbh	Percentile (0.1%, 1%, 5% and/or 10%) distribution of dbh
			Mean dbh in percentile class
2.5	Standard deviation of the tree heights	1 – tree height	Mean standard deviation of tree heights
2.6	Number of vertical layers	1 – number of vertical tree layers	A. Percent of one, two and multi-layered plots
			B. Average number of layers
2.7	Frequency distribution of standard deviation classes of dbh	1 – tree dbh	Mean standard deviation of tree dbh
2.8	Shannon index of tree species	1 – basal area by tree species	Mean Shannon index
3.1	Dominant age: mean age of the one hundred trees with the largest dbh on a per hectare basis	1 – tree dbh	A. Mean age
		2 – tree age	B. Proportion of plots older than half of the natural life span of the dominant tree species
		3 – natural life span (for B and C) 4 – even-aged high forest (for C)	C. Proportion of plots older than 120 years in even-aged high forest
3.2	Mean age	1 – tree age	A. Mean age
		2 – natural life span (for B and C)	B. Proportion of plots older than half of the natural life span of the dominant tree species
		3 – even-aged high forest (for C)	C. Proportion of plots older than 120 years in even-aged high forest

3.3	Weighted mean age	1 – tree age 2 – basal area by tree species 3 – natural life span (for B and C) 4 – even-aged high forest (for C)	A. Mean age B. Proportion of plots older than half of the natural life span of the dominant tree species C. Proportion of plots older than 120 years in even-aged high forest
3.4	Old trees: proportion of trees older than half of their natural life span	1 – tree age 2 – natural life span	Proportion of old trees
4.1	Deadwood volume by decay class, tree species, standing/laying	1 – deadwood volume 2 – decay classes 3 – tree species 4 – standing/laying	Average deadwood volume by decay class, tree species, standing/laying
5.1	Cover of ground vegetation	1 – vegetation ground cover classes	Proportion of plots in the different ground vegetation cover classes
5.2	Cover of shrub species	1 – shrub species ground cover classes	Proportion of plots in the different shrub cover classes
5.3	Presence/absence of shrub species	1 – presence/absence of shrub species	Proportion of shrub species on the total number of shrub species
5.4	Presence of shrub genus	1 – presence/absence of shrub genus	Proportion of shrub genus on the total number of shrub genus
5.5	Presence of life forms of ground vegetation	1 – presence/absence of life forms classes of ground vegetation	Proportion of plots with life forms of ground vegetation
6.1	Naturalness (N’)	1 – deadwood volume 2 – tree species dbh 3 – Shannon index of tree species	Frequency distribution of N’

6.2.1 Forest Categories

As reported in Sect. 5.2, the test objective was to classify the European NFI plots available in the common database with respect to the European Forest Types (EFT) system of nomenclature developed by the European Environmental Agency (EEA 2006). The indicator used is the forest category level of the EFT system.

Formal harmonization is necessary when a country adopts a local forest type system of nomenclature. However, harmonization is only possible when an unambiguous relationship between the classes of the two systems may be found. When a formal forest type system of nomenclature is not available locally, NFI plots may be classified according to the EFT system at category level on the basis of field data previously collected for other purposes. The geographical position of the plots can be used to obtain ecological characteristics such as soil type and climate conditions from other spatial products, while the plot observations and measurements furnish the information on tree species composition. On the basis of information from these sources, forest type can be determined and the forest category can be derived.

A common European system of forest classification, such as the EFT (EEA 2006), could be implemented as part of NFI field data acquisition activities. In this context, the operational implementation of the classification system would be similar to a standardization process. The EFT system could be slightly modified to accommodate local needs. A possible future development of the forest type approach could be the development of a system of nomenclature that includes both European and non-European forests.

6.2.2 Forest Structure

Using raw data for only a relatively few NFI variables, a large number of forest structure indicators may be estimated for evaluating horizontal, vertical and compositional forest diversity. The harmonization test was carried out by estimating eight different indicators: two are related to tree horizontal diversity, three are related to tree vertical diversity, and three are related to tree compositional diversity.

The data needed for the estimation of these indicators are common NFI data: tree species and their nativeness in the forest categories (Sect. 6.2.1), tree dbh, and tree height. Some of the indicators are based on estimation of traditional, aggregated plot-level statistics such as basal area and number of vertical layers. The harmonization effort first focused on making these raw data comparable across the countries using mostly reductive bridges and then estimating the indicators for the harmonized set of data. For vertical forest layering, an innovative method is presented and tested.

6.2.3 Forest Age

A standardized definition of tree age has been accepted at the international level, but a unique definition of forest age is still lacking. The development stage and the ecological potential value of a tree and, consequently, of a forest, are not linearly related to age. In our test, age was therefore standardized on the basis of the local expected natural life span of the tree species. The authors of Sect. 5.4 (Urs-Beat Brändli, Meinrad Abegg, and Jana Beranova) first proposed a draft table of natural life spans for several tree species in different environmental conditions.

Simple and ecologically meaningful forest age indicators may be estimated using tree age. Unfortunately, the ages of all the trees on a plot frequently are not acquired in the field by NFIs. Tree dbh may then be used as a simple proxy variable for tree age, but acquisition of tree age in the field is strongly recommended, at least for the dominant, largest, or oldest trees. Forest age indicators tested include the average age of trees, the average age of the trees weighted by basal area, and dominant age estimated as the mean age of the 100 trees with the largest diameters per hectare. Proportions of old trees and/or old plots were used to aggregate plot-level estimates of indicators by forest category or by geographical areas.

Following the harmonization test, dominant age estimated as the average age of the dominant trees of a stand, or its proxy, dominant dbh estimated as the average dbh of the dominant trees of a stand, was recommended as the indicator for forest age.

6.2.4 Deadwood

Deadwood volume is considered a classic forest biodiversity variable based on its adoption by multiple international frameworks. For our studies, the indicator tested was the total volume of deadwood classified by tree species, spatial positioning (lying/standing) and decay stage. The harmonization approach focused on estimating the deadwood volume of each single deadwood piece in accordance with different predefined reference definitions (Rondeux et al. submitted).

Harmonization of deadwood data is highly recommended before comparing aggregated statistics based on different deadwood definitions and sampling techniques. Harmonization of deadwood volume and decay stages is feasible, even if the necessary bridges are often complex. Harmonization of tree species and spatial positioning (lying/standing) of deadwood elements is not currently feasible because of lack of appropriate field data. The recommendation drafted for development and future modification of NFIs is to adopt local definitions of deadwood that are "harmonization ready" in the sense that they enable construction of simple bridges. Alternatively, an international standardized definition could be adopted.

6.2.5 Regeneration

Despite general agreement among NFIs regarding definitions adopted for assessing tree species regeneration, comparisons among countries using the regeneration data submitted for the common database were limited as a result of differences in NFI field observation and measurement protocols. Because of the limited data available and these protocol differences, an operational harmonization test was not possible for regeneration as was conducted for the other essential features. A consistent approach to harmonization should be conducted in the future to overcome these limitations.

6.2.6 Ground Vegetation

Harmonized estimation of indicators for ground vegetation was very problematic because of differences in definitions and methods adopted by NFIs for the acquisition of field data. The harmonization test focused on estimating two types of plot-level indicators: the first type was based on cover classes and included two indicators, one for ground vegetation and one for shrub life forms according to a harmonized system of nomenclature; the second type was based on presence/absence data and included three indicators, one each for shrub species, shrub genus and ground vegetation life-forms.

For each country or forest category, presence/absence observations for NFI plots represented in the common database were used to estimate two aggregated indicators. The first indicator was the percentages of the total numbers of shrub species and shrub genera appearing on harmonized lists that were observed on the plots. The second indicator was the percentages of NFI plots on which harmonized shrub cover classes were observed. Finally, plot-level data for the presence/absence of life-forms were aggregated as the proportions of plots with particular ground vegetation life-forms.

From an ecological perspective, these indicators probably have limited value. Nevertheless, we anticipate that this first experience will foster the understanding that forest biodiversity monitoring is currently severely limited by the lack of comparable forest vegetation data, with the exception of data for the tree component.

6.2.7 Naturalness

In Sect. 5.7, a new quantitative indicator of forest naturalness was proposed. Forest naturalness continues to lack a unique and commonly accepted meaning because different authors have interpreted the concept differently. Because of the complexity of the concept, the proposed indicator is also complex in the sense that it is based on three different sub-indicators: deadwood volume, tree species diversity quantified using the Shannon index, and tree size diversity quantified using the standard deviation of tree dbh. Even if the number and particular sub-indicators used change in the future, the general approach and the manner in which the data are aggregated to estimate a complex naturalness indicator is promising for future operational applications.

6.3 Lessons Learned

An initial and primary positive result achieved by COST Action E43 is the creation of a network of researchers interested in exploring the use and the possible improvement of NFIs. In WG3, this result was particularly relevant because the experts in forest monitoring and forest statistics had the opportunity to cooperate with forest ecologists, thereby merging and integrating different backgrounds, kinds and levels of expertise and investigation techniques. It is a new approach to use NFI data for monitoring forest biodiversity on a large-scale. We anticipate that this scientific community will continue this cooperation.

For forest categories and forest structure, we learned that harmonization is relatively simple mainly on the basis of reference definitions developed and reported by COST Action E43 (Tomppo et al. 2010). For European forest types, basic raw NFI data are usually sufficient to classify plots on the basis of the EFT (EEA 2006) system of nomenclature at category level even if specific reference definitions should be developed to harmonize such effort across different NFIs. For forest structure, we demonstrated that raw basic NFI data may easily be used to harmonize estimates of a wide range of structural and compositional indicators. For these two essential features, we expect that operational implementation of harmonized estimation will be feasible in the near future.

For forest age, naturalness, and deadwood, we proposed new innovative approaches for the harmonized estimation of indicators. For forest age and naturalness we achieved the objective of proposing and satisfactorily testing harmonized approaches to estimating quantitative indicators. These indicators are ready to be more widely and operationally tested. For deadwood, we estimated volumes of single deadwood elements using specific reference definitions. Although multiple reference definitions were tested within the framework of COST Action E43, a final reference definition with dimensional thresholds has not yet been fully accepted. We strongly recommend that the international scientific community commit to reaching a consensus on a reference definition and expect that the results of our tests will make important contributions to that effort.

Despite our successes, some questions are still open. The EFT system could be slightly modified to make a more direct connection with NFIs. Such modifications would facilitate more precise coverage of all European forest conditions. In addition, more consistent and objective rules for classifying plots would avoid possible subjective interpretation of the classification rules. If the EFT system is adopted by MCPFE for sustainable forest management reporting in Europe, then as soon as possible the plots should begin to be classified with respect to forest types based on direct field assessment rather than on a-posteriori computation of existing data. If the idea of reporting harmonized estimates of forest biodiversity indicators by forest type is widely accepted, then implementation of a global system of nomenclature could be considered.

Based on WG3 investigations and harmonization tests, reference proposals are now available for complete harmonized estimation of deadwood volume and for reporting by components (standing deadwood vs. lying coarse woody debris), decompositional stages, and tree species. However, the thresholds for the reference definitions still require agreement. The effects of differences in sampling designs and protocols on harmonization results obtained using deadwood volume models based on different reference definitions should also be investigated.

Estimates of some forest structure indicators may be affected by differences in sample plot configurations, particularly differences in plot dimensions. This problem may be at least partially overcome by recording the distances from sample trees to plot centres which can then be used to construct a reductive bridge based on the smallest sample plot among countries whose estimates are to be harmonized. However, accommodation for differences resulting from measurement of smaller trees only on smaller inner subplots of concentric circular plots as a means of reducing field crew workloads may still be necessary.

The indicators proposed for forest age, forest structure and naturalness are innovative but need to be operationally implemented in a test phase before proposing them as international references.

As previously noted, NFI information on ground vegetation and regeneration is insufficient with respect to both amount and comparability to be a good basis for harmonized estimation of biodiversity indicators.

6.4 Recommendations

The main unresolved issue is the impact of the number of trees per plot on the estimates of some of the indicators. For example, McRoberts et al. (2009) demonstrated that small plots and small numbers of trees per plot lead to less expected diversity in tree species. Plot-level sampling methods should therefore be harmonized with respect to plot dimensions, number of trees per plot and minimum dbh threshold. For these reasons we strongly recommend that NFIs adopt a minimum dbh threshold of 0 cm and also record the distances from trees to plot centres. These modifications will greatly facilitate the harmonization process regardless of the reference definitions adopted.

Harmonization of deadwood estimates is currently quite complex because of the large variability in definitions and field methods adopted by individual countries. If local sampling methods are valid, they should not be modified. For standing dead trees, we recommend adoption of a minimum dbh of 0 cm. For lying woody debris, we recommend length measurements and diameter measurements of both ends of deadwood pieces for purposes of obtaining at least the minimum and maximum diameters. We further recommend adoption of minimum diameter thresholds not larger than 10 cm and minimum length thresholds not larger than 1 m.

Application of the forest age indicators depends on the availability of age data for all trees measured and availability of local estimates of the natural life spans of tree species. We therefore recommend that countries develop locally optimized methods to acquire these data.

For ground vegetation and regeneration, the definitions and methods currently adopted by NFIs vary widely and merit serious investigation. We are convinced that a scientifically sound assessment of forest biodiversity cannot ignore the diversity of non-trees species. For biodiversity assessment, ground vegetation (herbs and shrubs) has as high potential relevance as trees. For these reasons, NFIs should also make a consistent investment in developing valid sampling procedures for these components of forest vegetation.

6.5 Conclusions

NFIs are the most complete source of information on forests, both in terms of number of sampling units and in terms of the information acquired in the field. We anticipate that in the future NFIs will also commit to becoming the most important source of information for monitoring trends in forest biodiversity, both in space and time.

The information currently available in most NFIs is appropriate for harmonized estimation of multiple forest biodiversity indicators, although mainly for forest structure and composition, forest age, deadwood, and naturalness.

Nevertheless, NFI programmes should be conscious of existing limitations and should continue the international collaboration initiated by COST Action E43 to improve their protocols in two important ways: (*i*) providing data and estimates based on local definitions and protocols that are optimized for harmonized estimation of indicators regardless of the international references that may be adopted; (*ii*) enlarging and enhancing acquisition of non-tree forest information. Regarding the latter point, some variables relevant for biodiversity should be immediately considered for the implementation into field protocols: (1) the botanical assessment of ground vegetation based on standard phytosociological methods such as Braun-Blanquet (1965), (2) the acquisition of information on habitat trees (Winter and Möller 2008), and (3) the acquisition of information on lichens and bryophytes. Further, other information related to forest biodiversity such as the assessment of birds and insects could be acquired for NFI sampling units.

Because the statistical estimators associated with NFI sampling designs are generally unbiased, harmonization with respect to them is usually not necessary for areal estimates of traditional forest variables such as basal area, growing stock volume, increment, and biomass. However, sample plot configurations and dimensions may affect estimates of indicators related to tree species composition and size. Thus, for purposes of orienting future NFIs toward biodiversity, we strongly recommend sampling units (or a subsample) be enlarged to at least 400 m^2.

Data acquired in accordance with these recommendations will enable future comprehensive and operational forest biodiversity monitoring programmes and will facilitate harmonized estimation of indicators that satisfy international forest biodiversity reporting requirements.

References

Braun-Blanquet, J. (1965). *Plant sociology: the study of plant communities*. London: Halner.

Convention on Biological Diversity (CBD) (2007). Article 7, Identification and monitoring. http://www.cbd.int/convention/convention.shtml. Accessed Apr 2008.

European Environment Agency (EEA) (2006). European forest types. Categories and types for sustainable forest management reporting and policy. Technical report 9/2006. European Environmental Agency. 111 pp.

European Environment Agency (EEA) (2007). Halting the loss of biodiversity by 2010: Proposal for a first set of indicators to monitor progress in Europe. Technical report number 11/2007. European Environmental Agency. 182 pp.

European Environment Agency (EEA) (2009). Progress towards the European 2010 biodiversity target. European Environmental Agency. Technical report number 4/2009. 52 pp.

Food and Agriculture Organization of the United Nations (FAO) (2005). Global resources assessment 2005. Progress towards sustainable forest management. FAO forestry paper 147. Rome. 323 p. http://www.fao.org/docrep/008/a0400e/a0400e00.htm. Accessed May 2008.

Hunter, M. L., Jr. (1990). *Wildlife, forests, and forestry: principles of managing forests for biological diversity* (370 pp). Englewood Cliffs: Prentice-Hall.

McRoberts, R., Tomppo, E., Schadauer, K., Vidal, C., Stahl, G., Chirici, G., Lanz, A., Cienciala, E., Winter, S., & Smith, B. (2009). Harmonizing national forest inventories. *Journal of Forestry, 107*, 179–187.

Ministerial Convention on the Protection of Forests in Europe (MCPFE) (2003). Improved Pan-European indicators for sustainable forest management as adopted by the MCPFE expert level meeting, 7–8 Oct. 2002, Vienna, Austria. Vienna: MCPFE Liaison Unit Vienna. 6 pp.

Montréal Process (2006). Criteria and indicators for the conservation and sustainable management of temperate and boreal forests. http://www.mpci.org/rep-pub/1995/santiago_e.html. Accessed Apr 2008.

Rondeux, J., Bertini, R., Bastrup-Birk, A., Corona, P., Latte, N., McRoberts, R. E., Ståhl, G., Winter, S., & Chirici, G. submitted. Assessing deadwood using harmonised national forest inventory data. *Forest Science.*

Ståhl, G., Cienciala, E., Chirici, G., Lanz, A., Vidal, C., Winter, S., McRoberts, R. E., Rondeux, J., Schadauer, K., & Tomppo, E., submitted. Bridging national and reference definitions for harmonising forest statistics. *Forest Science.*

Stokland, J. N., Eriksen, R., Tomter, S. M., Korhonen, K., Tomppo, E., Rajaniemi, S., Soderberg, U., Toet, H., & Riis-Nielsen, T. (2003). *Forest biodiversity indicators in the Nordic Countries* (Tema Nord). Copenhagen: Nordic Council of Ministers.

Tomppo, E., Gschwantner, T., Lawrence, M., & McRoberts, R. E. (Eds.). (2010). *National forest inventories. Pathways for common reporting*. Heidelberg: Springer.

Winter, S., & Möller, G. (2008). Microhabitats in lowland beech forests as monitoring tool for nature conservation. *Forest Ecology and Management, 255*, 1251–1261.

Winter, S., Chirici, G., McRoberts, R. E., Hauk, E., & Tomppo, E. (2008). Possibilities for harmonising national forest inventory data for use in forest biodiversity assessments. *Forestry, 81*, 33–44.

Index

G. Chirici et al. (eds.), *National Forest Inventories: Contributions to Forest Biodiversity Assessments*, Managing Forest Ecosystems 20,
DOI 10.1007/978-94-007-0482-4, © Springer Science+Business Media B.V. 2011

FSC
www.fsc.org
MIX
Papier | Fördert
gute Waldnutzung
FSC® C083411

Kolibri 360 GmbH
Ferdinand-Jühlke-Straße 7
99095 Erfurt, Deutschland
produktsicherheit@kolibri360.de